T0344397

Predictive Safety Analytics

Nearly all our safety data collection and reporting systems are backward-looking: incident reports; dashboards; compliance monitoring systems; and so on. This book shows how we can use safety data in a forward-looking, predictive sense. *Predictive Safety Analytics: Reducing Risk through Modeling and Machine Learning* contains real use cases where organizations have reduced incidents by employing predictive analytics to foresee and mitigate future risks. It discusses how Predictive Safety Analytics is an opportunity to break through the plateau problem where safety rate improvements have stagnated in many organizations. The book presents how the use of data, coupled with advanced analytical techniques, including machine learning, has become a proven and successful innovation. Emphasis is placed on how the book can "meet you where you are" by illuminating a path to get there, starting with simple data the organization likely already has. Highlights of the book are the real examples and case studies that will assist in generating thoughts and ideas for what might work for individual readers and how they can adapt the information to their particular situations.

This book is written for professionals and researchers in system reliability, risk and safety assessment, quality control, operational managers in selected industries, data scientists, and ML engineers. Students taking courses in these areas will also find this book of interest to them.

RELIABILITY, MAINTENANCE, AND SAFETY ENGINEERING: A PRACTICAL FIELD VIEW ON GETTING WORK DONE EFFECTIVELY

Series Editor: Robert J. Latino, Reliability Center, Inc., VA

This series will focus on the "been there, done that" concept in order to provide readers with experiences and related trade-off decisions that those in the field have to make daily, between production processes and costs no matter what the policy or procedure states. The books in this new series will offer tips and tricks from the field to help others navigate their work in the areas of Reliability, Maintenance and Safety. The concept of 'Work as Imagined' and 'Work as Done', as coined by Dr. Erik Hollnagel (an author of ours), this series will bridge the gap between the two perspectives to focus on books written by authors who work on the front-lines and provide trade-off decisions that those in the field have to make daily, between production pressures and costs . . . no matter what the policy or procedure states. The topics covered will include Root Cause Analysis (RCA), Reliability, Maintenance, Safety, Digital Transformation, Asset Management, Asset Performance Management, Predictive Analytics, Artificial Intelligence (AI), Industrial Internet of Things (IIoT), and Machine Learning (ML).

Lubrication Degradation
Getting into the Root Causes
Sanya Mathura and Robert J. Latino

Practical Root Cause Failure Analysis
Key Elements, Case Studies, and Common Equipment Failures
Randy Riddell

Predictive Safety Analytics
Reducing Risk through Modeling and Machine Learning
Robert Stevens

For more information on this series, please visit: https://www.routledge.com/ Reliability-Maintenance-and-Safety-Engineering-A-Practical-Field-View-on-Getting-Work-Done-Effectively/book-series/CRCRMSEGWDE

Predictive Safety Analytics
Reducing Risk through Modeling and Machine Learning

Robert Stevens

CRC Press
Taylor & Francis Group
Boca Raton London New York

CRC Press is an imprint of the
Taylor & Francis Group, an **informa** business

First edition published 2024
by CRC Press
2385 NW Executive Center Drive, Suite 320, Boca Raton FL 33431

and by CRC Press
4 Park Square, Milton Park, Abingdon, Oxon, OX14 4RN

CRC Press is an imprint of Taylor & Francis Group, LLC

© 2024 Robert Stevens

ISBN: 978-1-032-42438-5 (hbk)
ISBN: 978-1-032-42754-6 (pbk)
ISBN: 978-1-003-36415-3 (ebk)

DOI: 10.1201/9781003364153

Typeset in Times
by Apex CoVantage, LLC

Contents

Preface

Nearly all safety data collection and reporting systems are backward-looking: incident reports; dashboards; compliance monitoring systems, etc. In an age where buzz phrases such as "big data", "machine learning", and "AI" inundate us, how can we distinguish hype from reality? Are there ways to use safety data in a forward-looking, predictive sense?

The genesis of this book dates to 2010. A manager at a large railroad read the *Harvard Business Review* article "Competing on Analytics", authored by Tom Davenport (Davenport T. H., Competing on Analytics, 2006). He connected with the CEO of First Analytics, a firm that Tom co-founded, and asked how his company might improve its safety using analytics.

The railroad manager explained that safety was a top priority for the company and that it had improved considerably on this front, but it had gotten more difficult to keep improving. He said the company had already used some data to identify likely risks, but there was a lot more that could be explored. This led to a path of a proof of concept and subsequent multiple successful implementations of a predictive safety application across several operational functions (Davenport & Harris, Competing on Analytics: Updated, with a New Introduction: The New Science of Winning, 2017, pp. 112–114).

Just prior to this time, predictive analytics (later to include terms, like machine learning and artificial intelligence) was taking hold in commercial settings. But heretofore, the data and analytics approach to safety had never been attempted. And, to this day, there is not yet a widespread adoption of these tools for safety. This book aims to change this by making predictive safety analytics understandable and accessible.

This is not a theory book; it centers on real-world-use cases that demonstrate how predictive safety analytics can be applied. It shows that, regardless of a company's readiness with respect to the data they have, there is always a place to start.

Nor is this a technical book. Concepts such as statistical models and machine learning are explained in a non-technical introduction. Some of the software tools and technologies are also outlined so the reader may have familiarity with them. But no advanced degree in data science, statistics, or computer science is required.

As a part of the Reliability Maintenance, and Safety Engineering focus book series, this book offers tips and tricks from the field to help readers improve safety through the use of data – data they already have and data they may want to collect. Key takeaways include understanding how *predictive* analytics goes beyond *descriptive* analytics in assessing risk. The reader will hopefully realize how existing data may be able to identify and quantify risk factors, and to evaluate the effectiveness of safety programs and investments. But since no organization is fully mature in the adoption of these approaches, some of the simple use cases show that, while there are likely some gaps to address to get to a more predictive state, there are some steps to take right now to better leverage existing data.

Ultimately, the goal of this book is to engender a vision of being predictive, to intervene before incidents occur.

Author's Biography

Robert Stevens is part of the leadership team at First Analytics, a boutique analytical consulting firm. First Analytics designs and implements predictive analytics and machine-learning solutions. The firm services multiple industries with many applications. With an enabling engagement model, the firm teams up with its clients to build their in-house capabilities and systems. In his role as Vice President at First Analytics, Robert Stevens helps companies develop and execute programs to cultivate their analytics competency. He brings experience to bear, stemming from more than thirty years as an analytics professional, starting as an econometrician. His career has consisted of consulting, product development, client service, technical, and sales roles within software, consulting, and market research firms. Robert Stevens has participated in or led safety analytics implementations in the railroad, utility, oil and gas, and manufacturing industries. He has spoken on predictive safety analytics in venues such as National Safety Council congresses, an OSHA safety conference, human and organizational learning conferences, manufacturing forums, private analytics consortiums, and conferences and webinars sponsored by software companies.

Safety in Numbers

A Data-Driven Approach

1

THE BENEFITS OF SAFETY ANALYTICS

We all want our colleagues to return home safe and secure, with all limbs intact, with no injuries or trauma. A focus on personal wellness also has an impact on the business. Safety programs and processes, while protecting individuals, also bring benefits to an organization. Safety analytics can:

- Improve safety metrics such as OSHA[1] or FRA[2] reportable injury rates, DART (days away restricted time), lost time incidents, and near misses.
- Reduce lost productivity and improve operational metrics, like manufacturing OEE.[3]
- Reduce equipment damage.
- Reduce legal liability, litigation and settlement costs, and medical treatment payments.
- Reduce insurance premiums.
- Supply information to assist managers in safety-related coaching.
- Improve management/labor relations.
- Support managers in attaining their safety-related performance metrics.

---

[1] Occupational Safety and Health Administration, a United States regulatory agency
[2] Federal Railroad Administration, a United States regulatory agency
[3] Overall Equipment Effectiveness

DOI: 10.1201/9781003364153-1

- Aid safety professionals in measuring safety program effectiveness and in designing impactful policies, programs, and training.

These benefits delivered by safety analytics rely on data- and fact-based decision-making.

PROGRESS IN RECENT DECADES

Companies in all industries have strived, in the 21st century, to improve safety. And it has worked. Incident rates have been on a continual decline since 2003, as shown by data from the U.S. Bureau of Labor Statistics (Figure 1.1).

This is certainly good news. But, safety professionals are not satisfied, as demonstrated by the appearance of programs such as "zero incident goal" initiatives. The fact is, as we get better, it becomes harder to improve. This can be seen in the flattening out of the curves since 2015 and in the case of Days Away from Work, a reversal from 2020.

Rates have been improving for all industries

Nonfatal occupational injury and illness incidence rates by case type, private industry, 2003-2020

	2003	2004	2005	2006	2007	2008	2009	2010	2011	2012	2013	2014	2015	2016	2017	2018	2019	2020
Reportable Incident Rate	5.0	4.8	4.6	4.4	4.2	3.9	3.6	3.5	3.4	3.4	3.3	3.2	3.0	2.9	2.8	2.8	2.8	2.7
Cases with Days Away from Work	1.5	1.4	1.4	1.3	1.2	1.1	1.1	1.1	1.0	1.0	1.0	1.0	0.9	0.9	0.9	0.9	0.9	1.2

—Reportable Incident Rate   ····Cases with Days Away from Work

Incident rate = number of reportable incidents per 100 workers per year

FIGURE 1.1   Reportable incident rates in recent years

THE PLATEAU PROBLEM

This was the dilemma facing a manager when he approached the co-founder of First Analytics, Tom Davenport. Tom is a well-known author and speaker on the topic of advanced analytics, which includes such technologies as statistical (predictive) modeling, machine learning, and artificial intelligence. In what we might term "the plateau problem", the manager explained "that safety was a top priority for the company and that it had improved considerably on this front, but it got harder to keep improving" (Davenport & Harris, 2017, pp. 112–114). Furthermore, he said "The company had already used some data to identify likely risks, but there was a lot more that could be explored."

The answer to his dilemma was data. Not just data alone, but data in conjunction with predictive analytics. The result for his company was breaking through the plateau. They experienced four consecutive record years, reducing reportable injury rates by 28%. While they previously were middle-of-the-pack in their industry, after implementing analytics, they led the industry, with a rate 26% better than their closest peer.

While this company's solution was operational – providing risk assessments at the employee/day/shift level, which required a lot of data – there are many possible higher-level, strategic applications that can be established in short order and without onerous data requirements.

This book illustrates, with real-world examples, how companies can use Predictive Safety Analytics on top of data they already have to further improve employee and public safety.

THE FOUR PS OF SAFETY

Analytical tools can be pointed toward a wide variety of safety issues. While this book shows some examples, they are in no way comprehensive. To begin to paint the picture of the breadth of applications available, it is useful to classify them. A particular application area may not be covered in the book, but this framework should inspire ideas as to how data and analytics can be employed in your context.

Marketing is known for the "four Ps of marketing": **p**roduct, **p**rice, **p**lace, and **p**romotion. Borrowing from marketers, we have **P**eople, **P**laces, **P**rocesses, and **P**redictive Analytics. Figure 1.2 shows where we can focus predictive analytics within the first three Ps and what the desired outcome would be as a result.

People

Ultimately, the goal of safety programs is to protect individuals, both employees and the public. But the way this is done can both be strategic and tactical.

Focal Points and Outcomes

Focal Point	⟶	Outcomes

People
- Employees
- Customers
- The public
- Managers

- Reduce incidents, accidents & injuries
- Care for your employees
- Protect the public
- Coach managers

Places
- Facilities
- Environments
- Fleet
- Equipment

- Reduce lost productivity
- Improve facility metrics
- Discover where to invest in equipment
- Alert to dangerous contexts

Processes
- Policies
- Programs
- Training
- Compliance

- Measure safety program effectiveness
- Design impactful training
- Strengthen policies
- Foster good employee relations

FIGURE 1.2   The four 'Ps' of Safety

EHS[4] professionals develop programs and policies to protect employees and the public in a broad, general sense. These programs and policies represent their safety strategy.

Tactical applications, on the other hand, focus on a particular person within a given context. For example, an employee working a specific shift. This is also known as operational safety analytics. The availability and richness of highly granular data now allows us to make risk predictions for specific individuals working at a particular time and location. The risk assessment is also tied to the function they are performing or the assigned work order.

The goal of both strategic and operational safety analytics is to reduce incidents. Most commonly, these are measured by such metrics as TRIR (total recordable incident rate) or DART (days away from work, restrictions, transfers). In the United States, the Occupational Safety and Health Administration (OSHA) provides standards for these metrics. Industries and countries vary on the actual kinds, names, and definitions of metrics, but all are based on incident rates of some kind. Even if not required to be reported to a regulatory body, most companies track things such as:

- Reportable and non-reportable incidents and accidents
- Injuries requiring medical treatment
- Incidents requiring minor first aid
- Fatalities

---

[4] Environment, Health and Safety, or sometimes, HSE

- Lost time incidents
- Near misses and close calls.

The primary outcome of people-level analytics is to reduce incidents, accidents, and injuries by starting at the individual level. These tools also equip managers to have positive (never punitive) conversations with at-risk individuals who rise to the top of the safety risk list. These conversations are based on data and facts.

In reality, nearly all companies are not ready to employ this atomic level of risk mitigation. But that does not mean they cannot get started. As will be seen in the case studies, there are many areas where analytics can start making an impact without operating at the individual level.

Places

Places are the context within which incidents may occur. These can be facilities, like manufacturing plants, distribution centers, oil rigs, construction sites, cellphone towers, transmission substations, etc. And many job tasks are performed outside of a physical facility, such as operating a train locomotive. Places can also include the surrounding environment, such as the weather, which is very often a contributor to risk. Finally, many companies maintain fleets, where drivers operate trucks or other vehicles, putting themselves and the public at risk.

Data collected about places, fleets, and the environment allow us to quantify the risk related to those assets versus the risk associated with an employee. They allow us to discover which facilities or locations are the most at risk. They also can show us what machinery is more dangerous.

Besides the obvious impact of reducing incident rates, other outcomes of analytics focused on the Place level include:

- Reduction of lost productivity and down time
- Improvement of facility metrics, such as plant OEE (overall equipment effectiveness)
- Discover where to invest in equipment
- General alerting to dangerous contexts, such as severe weather.

Processes

Some industries, particularly the oil and gas and chemical industries, are primarily oriented toward so-called process safety. There are established elaborate standards and protocols that we won't detail here. Suffice it to say, if a process can be measured, it can often be modeled.

Beyond operational process, there are other aspects of a company's safety culture that fall under the category of processes. These include things like policies, programs, and training. The impact of each of these can also be measured, as several case studies will show. In this fashion, companies can determine which of these have been effective in the past and which ones have not. This helps in refining the approach to bring an overall safety culture to a company.

The outcome of analytics focused on processes includes:

- Measuring safety program effectiveness
- Designing impactful training
- Strengthening policies
- Fostering good employee relations.

Predictive Analytics

The fourth 'P' is predictive analytics. This is the math and data that connects the elements of the three other Ps with the outcomes. In other business operations and contexts, apart from safety, predictive analytics usually has a small, or even singular, use to connect inputs to a decision and an outcome. In contrast, predictive safety analytics has many applications, targets, decisions, and outcomes. Some of these will be illustrated in the used cases that follow.

Analytics Defined

2

WHAT WE MEAN BY "ANALYTICS"

Analytics can mean many things to many people. Broadly speaking, any kind of information derived from data can be seen as analytics. For our purposes, we distinguish, as do many in the data science profession, between *descriptive, predictive,* and *prescriptive* analytics. These types, along with the sample questions they address, are depicted in Figure 2.1.

This classification was originally established by Tom Davenport in the book *Competing on Analytics* (Davenport T. H., Competing on Analytics, 2006). He and others have since elaborated on and extended this framework. They follow an increasing hierarchy of value delivered. As we get more predictive and more prescriptive, the greater the degree of intelligence is provided.

Descriptive Analytics involves gathering, describing, and depicting data. Conveyed via static reports and dashboards, this kind of analytics is backward-looking. It includes things like standard reports, visual analytics, querying tools, and alerts based on business rules. While useful for status reporting, it does not tell us much about why the results are happening or what might happen in the future.

Predictive Analytics uses data from the past to understand relationships among the different variables. The intent is to predict the likelihood of a phenomenon happening on the basis of those relationships. Techniques such as statistical modeling or machine learning are used to find those relationships. The models help us understand cause and effect and answer the question, "What happens next?"

Prescriptive Analytics starts with a predictive model and suggests a course of action. Tools such as the "What if" scenario analysis and experimental

DOI: 10.1201/9781003364153-2

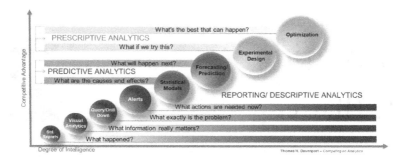

FIGURE 2.1 Types of analytics

design answer the question, "What if we try this?" And mathematical optimization answers, "What is the best that can happen?" While prescriptive analytics is very closely related to predictive analytics, in this book, we will mostly emphasize predictive analytics. This is mostly because nearly all companies still need to take the first step to go from descriptive to predictive analytics. This is the aim of this book.

NON-TECHNICAL INTRODUCTION TO MODELS AND METHODS

The reader less interested in models and technology can skip ahead to chapter 3. This chapter can serve as a reference, if needed.

Models are at the heart of predictive analytics. But what defines a model is not always clear. There are many terms associated with models and modeling, and those, too, can add to the confusion. Even amongst quantitative professionals, there is sometimes debate and ambiguity with definitions. In fact, even the professional titles of Statistician and Data Scientist spark debate.

Quantitative Professionals

Before describing models and methods, we will talk about the people who build, deploy, and interpret them. Not desiring to jump into the definitional debate, we simply refer to the definitions provided by the *United States Bureau of Labor Statistics in the Standard Occupational Classification Manual* (Bureau of Labor Statistics, 2018).

15–2041 Statisticians

Develop or apply mathematical or statistical theory and methods to collect, organize, interpret, and summarize numerical data to provide usable information. May specialize in fields such as biostatistics, agricultural statistics, business statistics, or economic statistics. Includes mathematical and survey statisticians. Excludes "Survey Researchers".

15–2051 Data Scientists

Develop and implement a set of techniques or analytics applications to transform raw data into meaningful information using data-oriented programming languages and visualization software. Apply data mining, data modeling, natural language processing, and machine learning to extract and analyze information from large structured and unstructured datasets. Visualize, interpret, and report data findings. May create dynamic data reports. Excludes "Statisticians", "Cartographers and Photogrammetrists", and "Health Information Technologists and Medical Registrars".

To make matters worse, there are now also data engineers, closely associated with data scientists. The BLS has yet to add this profession to the classification manual. They perform tasks related to the usually complex preparation of data needed for models. And machine-learning architects are those who design end-to-end machine-learning systems, including how the data flows through an application.

For prescriptive analytics applications, there are industrial engineers – professionals with degrees in operations research or mathematical optimization. Our focus in this book is mainly on predictive analytics.

To undertake predictive safety analytics, you need some of these professionals either on staff or through a firm that provides these services.

Something often heard is, "We need to hire a data scientist" without much thought about what that means. The reply is, "What kind of data scientist?" The field is so broad now that, while there are some good generalists, there is a lot of specialization among practitioners. This is like saying, "I need to see a doctor." Well, what kind of doctor? A general practitioner? A cardiologist? An oncologist? A neurologist?

Data scientists may be especially experienced in marketing analytics, supply chain, predictive maintenance, machine vision, natural language processing, and so on. Layer on top of that a particular focus on a specific industry. So, it is not as easy as just "hiring a data scientist" – it takes some thinking to lay out the specifications for what you need. But this should not be discouraging. It merely means you need to give some thought to outlining the specific needs your company. The following discussion should help you have a basic level of familiarity with the tools and technologies involved, without needing to have a deep comprehension of how they work. Having this high-level understanding makes one a better consumer of models.

Data Science Pipeline

Data scientists speak of their workflow in terms of a pipeline. In fact, many modern software tools for data science build upon this concept to manage the tasks associated with building models. There is a myriad of ways pipelines are put forth, and no one scheme rules them all. A very simple representation of a pipeline is shown in Figure 2.2.

The steps in the pipeline flow are the following:

Source data includes a wide variety of data sources, with varying content and in different formats. Examples of these are discussed further in the chapter titled, The Safety Data Repository. The data can come directly from an existing safety management system (SMS) or other operational databases. It may also include third-party and external data. The formats can include formal database structures such as SQL Server, Excel files, or flat (e.g., comma-segmented values) files. The data engineer has tools to extract and load data from each of these sources.

Import and process is also known as data ingestion and integration, or extract, transform, and load (ETL). Sometimes the euphemisms "data munging" or "data wrangling" are used. A part of this step includes reporting on data anomalies. Most data contain flaws in that there are missing values, extreme values, or infeasible values. The processing step includes applying remedies for these situations.

Data Science Pipeline

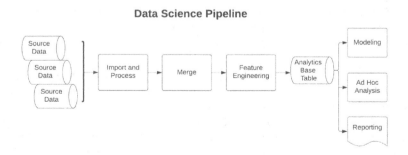

FIGURE 2.2   A simple data science pipeline

The remedies themselves can be time-consuming and complicated in and of themselves.

Merge is where the data is combined. Data sets are joined using key variables. For example, an employee ID number in an incident reporting system would be used to join with human resource data about the employee using that same ID number as the key. Merging poses some very daunting challenges that require ingenuity to overcome. For example:

- Many key variables (ID variables) have different schemes across databases; a key variable in one data source may not correspond to key variables in others.
- Some databases have no key variables at all, or at least key variables that correspond to their counterparts in other data sources.
- Data sources come with different hierarchies that need to be reconciled. For example, an employee may report into a functional group, within a district, and then within a management region. Other systems may have different hierarchy schemes that represent the same element.
- Furthermore, there are multiple hierarchies: people; geography; assets; time; etc.
- Time granularity and intervals of measurement can vary from by-the-second, in the case of sensor/event recorder/telematics data, all the way up to the annual level, in the case of employee engagement surveys.

Feature Engineering is the phrase data scientists use to describe special transformations and derivatives of the variables in the raw data. The machine learning community uses the term "feature" to mean the same thing that the statistics community calls an "independent variable". A database professional may simply call them "columns".

Feature engineering attempts to create representations of the raw data that enable statistical models or machine learning algorithms to find more definitive and interesting patterns than using the raw data alone. For example, the time clock data which measures shifts worked can be expanded to create variables regarding the number hours worked in the previous three, seven, fourteen, or thirty days (as examples). These become a proxy for potential fatigue. Examples of transformations include:

- The sum or average of a metric in the past N days (with multiple Ns)
- The number of days since . . .
- Leading/lagging transformations of variables (note that some algorithms can estimate time-shifted relationships automatically).

- Binning: Creation of a number of "buckets" of variables whose values fall within ranges
- Mathematical transformations of variables, such as taking the natural logarithm of a number.

As good as machine-learning algorithms have become, in terms of pattern recognition, much of the strength of a predictive system comes from careful feature engineering. Thus, much time is spent in this exercise, usually much more time than developing the algorithm itself. The <u>Analytics Base Table</u> will be described in the chapter on the Safety Data Repository.

Models

For disambiguation, we consider a model to be a mathematical method to either (1) make a prediction or (2) to quantify a relationship between variables. Some types of models do both tasks at the same time. A model is not simple arithmetic that fits into a spreadsheet. Models attempt to uncover relationships between variables or patterns that can be quantified with model parameters or other mathematical structures. Once the values of those parameters are known, they can be used to perform one or both of the two tasks. The act of building a model is to use data to estimate the values of those parameters.

Prediction is sometimes called forecasting, although there can be differences in the way the two terms are used. Forecasting is most often used in the future-looking sense. For example, "How does this trend project in the coming weeks?" Prediction more commonly determines the likelihood, or propensity, of a given event to occur within a specific context. For example, "What is the propensity that this worker will be involved in an accident?"

In the latter case, when predictions are to be made, the data scientist will call this "scoring". In other words, given a new set of data, what probability, or "score", would the model estimate, given the inputted values from the data. Alternatively, the word "inference" is used to mean "scoring".

Sometimes it is useful to understand the relationship between variables. Safety professionals often speak of "leading indicators". In this sense, there is thought to be a correlation across time between the two variables. Knowing that correlation helps understand what might happen with one variable as another one changes. In modeling terminology, quantifying these relationships can be called causal analysis or inference. The variable of focus is called the dependent, or target, variable. The variables thought to be correlated to this variable are called independent variables, causal variables, or features.

Admittedly the terminology and vocabulary can be difficult to grasp. A problem we have in the analytics profession is that different disciplines have addressed the same problems with their particular approaches and from their differing backgrounds. Statisticians and computer scientists both have methods to solve these problems. It can be confusing when a statistician describes their *regression model* and its *independent variables* while the data scientist, coming from the computer science ranks, talks about their *supervised learning algorithm* and its *features*. These phrases describe the same concepts, and, in fact, the actual methods can be identical!

Abbreviated Taxonomy of Models

The number of potential methods to tackle both the prediction and inference tasks are innumerable. But since users of predictive safety analytics models will hear a variety of names thrown about, it is useful to provide a framework and classification that generalizes the approaches. The following is in no way comprehensive, but provides some basis for understanding, at a high level, what each of these more common methods are suited for. It is not intended as a primer for these models, but instead as a reference for when one encounters these terms.

Speaking in general terms, most models are categorized as *supervised learning* or *unsupervised learning* techniques.

Supervised Learning. In supervised learning, the machine is taught by example. Examples of the outcomes along with the inputs are provided, and the machine uses its logic to determine the relationship between the inputs and outputs to make predictions. Common supervised learning techniques are described below.

- Simple Linear Regression. This is the model that is taught in every elementary statistics course. It supposes that the dependent, or target, variable, **Y,** is a function of the independent, or input, variables, **X,** and some unknown, but to-be-estimated parameters, β (the Betas). It is called linear regression in that the relationship between the parameters and the inputs is simply multiplicative: $Y = \beta X$. When this relationship is sketched out, as **X** increases, a straight line represents the response of **Y**. Regression has the appeal of not only being predictive, but it can quantify the relationship between the independent and dependent variables in an easily interpretable way.
- Variations of linear regression. Sometimes the target variable is thought to follow a response pattern that may not be strictly linear. Yet, within this same linear regression framework, many of those

phenomena can be accommodated through special transformations of the variables. So-called *polynomial regression* allows for curvilinear relationships. And, taking the natural log of the variables results in a *log-log* or *multiplicative* model. In this model, the parameters can be interpreted as elasticities, or the percent change expected in the target variable resulting from a 1% change in the input variable.

- Generalized Linear Models (GLM). The regression methods presented previously were described in the context of *continuous* target variables. That is, variables that can take on an infinite number of values within a range and typically result in measuring something. So-called *discrete* variables take on a limited, distinct number of values within a range and are typically the result of counting something. For example, the number of incidents in an operating region is discrete in that it can take on only the values of 1, 2, 3, etc. In generalized linear models, the basic simple linear foundation is still in place, with special transformations of the variables used to account for the discrete nature of some of the variables.

- Logistic Regression. This is a special case of a generalized linear model, where the target variable is discrete. These are common and very useful in safety analytics in that they can be used to predict the probability of an incident occurring. In this case, the target variable is binary: Did the event occur or not? This binary (0/1) response is the target against which the model estimates the relationship of the potential input variables. The logistic regression model also has the appeal of "explainability", in that the model parameters can be used to quantify how each input increases or decreases the odds of an event occurring. Even with the emergence of more sophisticated machine-learning approaches, this method performs very well and should always be considered a powerful tool in the data scientist's toolbox.

- Categorical and Ordinal Regression. These models also fall within the generalized linear model framework and are an extension of logistic regression. For categorical regression, the target variable can take on more than one discrete value. For example, the type of incident could be a first aid event, near miss, lost time, injury, etc. The model treats these probabilistic relationships, with the input variable, differently for each type but keeps it within a single model. Ordinal regression extends the categorical model by imposing a ranking structure on the categories. For example, the severity of an injury may follow a specific, graduated scale.

- Decision Trees. Variations of this method can be used to predict both continuous and discrete target variables. They are popular because of their ease in interpretation, although they tend to not perform as well in validation tests of their predictions. The structure of this model is represented by a hierarchical, treelike structure, with branches extending from various "decision" nodes. For example, the top node may be, "if snowing", the second node may be "if performing task A", then the final outcome node may be "probability of an injury = X".

- Random Forests and XGBoost. These approaches extend the single decision-tree model to combine many decisions-tree models. This can alleviate the prediction performance issue mentioned previously when a single tree is used. In a sense, this jury of opinion of the models can boost the performance (in fact, there is an approach called Boosted Trees, of which ADAboost is one example). There are many varieties of the tree ensemble approach – too many to mention here. And it is worth noting that, while the ensemble approach increases predictive accuracy, it often does so at the cost of less interpretability. One approach that has recently demonstrated great success recently is XGBoost (Extreme Gradient Boosting), and a similar approach is CatBoost, so they are mentioned explicitly here.

- Neural Networks. This is one of the original machine-learning approaches to prediction. Its roots go back many decades. It was the first attempt by cognitive scientists to replicate how the human brain functions. An input is processed through a network of connected neurons, ultimately producing an output, similar to how the brain is structured. As with other techniques, there are many variations and derivatives, too many to list here. But some names you will hear are multilayer perceptrons, convolutional neural networks, recurrent neural networks, deep learning, LSTMs, and radial basis functions. Neural networks traditionally suffer from having little to no interpretability. They are the most "black box" of the methodologies, where it is difficult if not impossible to understand how they come up with their predictions. This has been changing in recent years, as methods applied against those and other models, such as *Shapley Values*, can be applied to them to quantify variable importance. Most neural networks also require very large datasets. For the most part, neural networks have not met with much success in safety analytics. But innovation in neural networks and other methodologies is rapid, and one should never dismiss out-of-hand a suggested approach.

- Time Series Models. These are models that look to project or forecast trends and patterns into the future. In that sense, they are predictive but are used in a different way. See the previous discussion about the distinction between prediction and forecasting. Traditional approaches to time series forecasting go by the names ARIMA, or Exponential Smoothing, models. But several of the other methods discussed, such as regression, and certain types of neural networks can be used for forecasting.
- Others. Less common supervised learning algorithms you may come across include Naïve Bayes and Support Vector Machines.

Unsupervised Learning. In unsupervised learning, the machine is looking for patterns and not necessarily making predictions. Since this book focuses more on predictive techniques, we give unsupervised learning techniques less consideration. Common unsupervised learning applications are described below.

- Clustering. This is a broad approach to grouping based on similarities or differences. For example, establishing cohorts of employees whose profiles are similar based on training, tenure, skills, etc. Clustering is often used for producing insights, such as reporting historical incident rates by group. But clustering can also be performed in advance of predictive modeling to help increase the predictive accuracy or to produce predictions by group profile. The names of some of these methods included hierarchical clustering (including Ward's method and average linkage), K-means clustering (the most common and easy to implement solution), and more sophisticated probabilistic clustering, such as Gaussian mixture models. Many of these are statistical approaches, but techniques coming from the machine-learning community also have applicability.
- Anomaly Detection. This approach attempts to find unusual instances or patterns in data. Classic approaches include traditional statistical outlier identification (mean plus standard deviations). But a wide variety of models and methods can be employed to isolate those instances that are outliers from what a model would expect.
- Association Rules. Also known as affinity analysis, these methods look for combinations of one or more conditions that are more common than what might be expected. For example, if you combine employee #1 and employee #2 into a team to perform a task, there may have been a history of elevated risk with them working together in the past. Common methods are the Apriori algorithm and FP-Growth.

SOFTWARE AND TECHNOLOGY

Safety Management Systems

Generally speaking, a safety management system is a set of objectives and processes to maintain safe operations. But technology is used to reflect these systems. Our discussion relates to the technological aspects of a Safety Management System (SMS), and we will be using this term in the technological sense.

These systems exist to record incidents, assure compliance, measure performance, etc. They are software and web-based tools that include querying and reporting on safety data. There is often a dashboard associated that presents stakeholders with a high-level view of recent safety performance.

The underlying data in these systems can vary widely in their formats and content. The appeal of an SMS is to collect these data into one place and offer them up to users. But, as we will see, even these systems are limited in terms of being able to use all kinds of data that may be predictive of risk. When we discuss Predictive Safety Analytics, we do not suggest replacing an SMS; rather we talk of *augmenting* an SMS and integrating with it. No predictive safety analytics project has ever suggested that an organization should throw away their investment in an SMS. Rather, PSA builds upon that investment.

The main limitation of an SMS is that they are backward-looking, reporting on historical trends. There are very little, and almost always no, forward-looking aspects of these tools, which are things like trend projections or predictive models of risk using the aforementioned tools. As set out at the preface of this book, we encourage a move from backward-looking reports and status reporting to a forward-looking approach with predictive tools.

Some common SMS vendors include Enablon, Intelex, and Synergy. Larger enterprise software vendors, such as SAP, offer SMS modules. Many SMS systems are developed and maintained in-house.

It is beyond the scope of this book to enumerate and evaluate SMS tools. But we do point out that, when evaluating vendors, it is very important to assess their openness. That is, to what extent do you own the data that is put into the system. Can you access it for other purposes at its lowest level? Are there interfaces that make it easy for data to be extracted? This is critical for augmenting and integrating advanced analytics, along with some of the tools mentioned next.

Analytics and Machine-Learning Tools

This brief section provides a simple list of the more common tools and frameworks you may hear of. You may not be a carpenter or know much about carpentry, but you know about the basic tools they use. Thus it is the same with data science and predictive analytics. Though you may not be doing the work, it helps to understand the tools involved when talking with those who do.

Some tools, particularly the more recent ones, are open source. This generally means they are free to download and use, although there are variations of open-source licenses that do have restrictions. If you plan to commercialize an application in these tools, make sure you check with the license restrictions, as "free" may mean free to use, but not to distribute applications with them.

Also, "free" may not really be free in the sense of a total cost of ownership. You may hear the phrase, "What is the TCO?" Open-source software can notoriously be tricky to implement and maintain. Although there are communities that may provide advice, there is no vendor to call on to provide technical support. And there is nobody accountable to guarantee the function of the software and whether it is doing the expected task correctly, with regard to the math, for example. And since packages can be updated frequently by anyone who contributes, your implementation of that package may break unexpectedly upon a new update of the software.

There are software companies that attempt to bridge the gap between open source and proprietary software to address some of these problems. They offer stable versions of the open-source code, which they may have also optimized. And, they usually offer support for a fee.

General Analytics-Oriented Languages

- Python is a broad programming language with strengths in manipulating data. A large variety of "libraries" are available that are purpose-driven, such as implementing a specific machine-learning algorithm or a collection of algorithms. Related to libraries, and often synonymous with them are "frameworks". These work as add-ins or wrappers around Python to do modeling. Some of the more common ones are Pandas, Scikit-Learn, Numpy, Tensorflow, Keras, PyTorch, Theano, and LightGBM. This ecosystem of tools rapidly evolves. Python has emerged in recent years as the toolset of choice for data scientists. But this area evolves quickly. Python is open source.
- R is an open-source software package that traces back to the 1990s. It traditionally had been a statistical modeling programming language, and thus primarily used by statisticians (still to this day).

It was a favorable alternative to proprietary software until recently, as Python has emerged with its dominance. There used to be a firm distinction between R and Python, the former being statistical modeling focused, while the latter incorporating the newer machine-learning algorithms. Those lines, however, have blurred, with many statistical libraries available in Python, and conversely, some machine-learning packages available in R.

Analytical Software Packages and Platforms

These are collections of tools with various functionality. As larger solutions, most of these are proprietary and have annual license fees, based on the number of users or the size of the hardware where installed. Open-source packages have taken away market share from these vendors, but they have responded by positioning themselves as platforms that also incorporate the open-source tools.

While it is tempting to think one is saving money by focusing exclusively on open-source software, one must consider the total cost of ownership ramifications discussed previously. These proprietary vendors offer a complete, integrated, and common framework for doing analytics throughout the pipeline.

Some of the more prominent vendors include SAS (market pioneer and leader), Dataiku, Alteryx, DataRobot, Knime, RapidMiner, IBM SPSS, and H2o.ai.

The software landscape can be bewildering and ephemeral. Industry analysts, such as Gartner, IDC, and Forrester, regularly issue reports evaluating these solutions to help with selection decision-making.

Analytics in the Cloud

You may have heard your company's IT department refer to "the cloud". This means that the physical hardware is not owned and maintained by the company "on premises". Rather, the data storage, processing, and computing occurs on servers maintained in data centers by a cloud outsourcing company. You access these systems through secure internet connections.

The three major cloud providers are Azure by Microsoft, Google Cloud Platform (GCP), and Amazon Web Services (AWS) from Amazon. In fact, your SMS system, whether proprietary or in-house developed, may already be hosted in the cloud. This was not always true of safety-related data, due to security concerns and the sensitivity of that data. But this has been changing, as cloud platforms often have better security than on-premises systems that tend to be more susceptible to data breaches.

Each of the three major cloud vendors offer data processing, feature engineering, and modeling tools for the data scientist. And since they offer data storage, data processing, query and reporting tools, and business intelligence tools, they sometimes have appeal as an integrated platform, rather than as a collection of one-off tools and more complicated data flows. Examples include Microsoft Azure Machine Learning, Amazon SageMaker, and Vertex.AI from Google. Other analytical tools exist, including automated machine learning, which automates many of the tasks of the data scientist, and data/feature engineering tools geared toward analytical modeling.

There are cross-cloud platform providers, such as Databricks, based on an open-source data technology called Spark, that focuses on large-scale data processing. They incorporate many of the analytical tools previously mentioned to make analytics at a large scale possible.

The Safety
Data
Repository

3

THE CHALLENGES OF
SAFETY-RELATED DATA

Of all the data-generating and data-consuming parts of an organization, perhaps none is as complex as those related to safety operations. There are potentially a wide variety of data types and formats, disparate structures and granularities derived from dozens of sources with different contents.

A *Safety Data Repository* is a way to organize this data. It is curated data for safety analytics, such as reporting and visualization, ad hoc analyses, and importantly, predictive analytics. Its features and benefits include:

- Data lake and database schemas tailored for safety analytics
- Integrates with but does not replace a safety management system (SMS)
- Draws from dozens of data sources in varied formats
- Deals with various levels of aggregation
- Distills text from reports into structured data
- Uses special data transformation and joining methods for predictive modeling and machine learning.

DOI: 10.1201/9781003364153-3

PRIMARY AND ANCILLARY
DATA SOURCES

Collected and collated data in a safety data repository can be from primary safety-oriented sources, as well as secondary or ancillary sources that may relate to safety. Primary sources are typically housed in a Safety Management System, whether proprietary or home-built. They contain data on:

- **Targets**: incidents; injuries; near misses; lost time incidents; DART; first aid; OSHA reportables, process safety events; etc.
- **Leading Indicators**: observations; audits; inspections; behavioral-based safety; compliance; etc.

Ancillary sources can be anything that is related to safety, including those with potential causal relationships or those which provide contextual background. They come from a variety of operational systems. Examples include:

- **Work Shift**: scheduling, time keeping, function; location; craft, tasks, teams.
- **HR/Employee:** attendance; PTO; tenure; demographics; performance reviews; engagement surveys; training; knowledge tests; certifications; fatigue scores; commute time, drug/alcohol tests; discipline; cohort; union membership; management.
- **Assets**: site characteristics; maintenance; downtime; equipment changeouts; vehicle telematics; event recorders; sensors; video.
- **Operations**: processes; rules; policies; PPE use; business velocity.
- **Corporate**: training programs; safety campaigns; legal.
- **External**: weather; macroeconomic.

This list, while extensive, is not comprehensive. And one need not have all of these in place to begin analytics. Rather, it is best to take an incremental approach to building the data repository over time. Some of the steps will be described in Chapter 5 to help you at the outset of your analytics journey.

THE ANALYTICS BASE TABLE

In a previous section on the data science pipeline, we describe general steps for importing and processing, merging, and performing feature engineering on the data. For specific analytics applications, the final data is collected into a so-called Analytics Base Table. It is often referred to as "the ABT". This is the final data, staged for use with a machine-learning or predictive model, or other data science algorithms.

A depiction of a real-world pipeline flow to create the ABT is shown in Figure 3.1. This flow has been implemented with many clients.[1]

The ABT is a single table with (usually very many) rows and columns (sometimes dozens but often hundreds). The columns represent transformed (through feature engineering) versions of the original raw data. There can be many realizations of a single-source variable. For example, a raw variable that records the number of hours worked can be transformed to show how many hours the employee worked in the past 3, 7, 14, and 30 days. These three all go on the row record for the current shift.

Even a small number of source variables, say a dozen, can turn into hundreds of variables in the ABT. To organize these, they should be put into categories. Example categories are schedule, attendance, job function/title, training, tenure/experience, supervision, team membership, discipline, working location, calendar (seasonality (month, day)), operational environment, physical environment (climate), individual profiles (demographics).

In the process of discovering and evaluating source variables, a factor taxonomy is built. This is a classification scheme that assigns variables to categories and subcategories. Not only is this a way of systematically organizing and preparing the data through feature engineering, but it helps to develop the thinking and discussion about hypotheses that need testing. A factor, like coming off vacation, should be assigned to a category, possibly a subcategory, and suggested hypotheses provided. The process of building this taxonomy will reveal insights and issues with the source data itself, so this is a valuable exercise in providing feedback to the information technology team.

Finally, good practice dictates the creation of a data dictionary. For every variable in the ABT, there is metadata about that variable, including such elements as:

- the variable short name for used in the modeling;
- a longer description of the variable;

---

[1] Credit to Bill Qualls, Chief Data Munging Officer of First Analytics

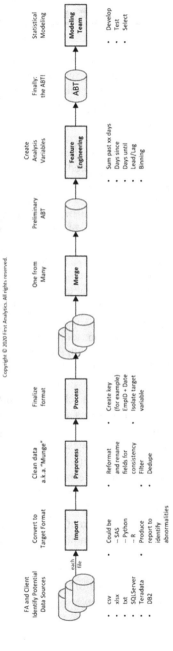

Creating the Analytics Base Table (ABT)

FIGURE 3.1   The Analytics Base Table (ABT)

- the factor category (and possibly subcategory) it belongs to;
- the source data from which it comes, including possibly the source database table and column name;
- the date range of the data availability (there is often varying lengths of historical data available, however, an ABT usually harmonizes everything to a unified time period);
- information on how often the data is refreshed;
- the owner (department or named individual) of the data;
- notes.

The analytics base table is the core of the predictive analytics system. A lot of time and much care goes into its construction. Peter Norvig, Google's Chief Scientist, famously said, "We don't have better algorithms. We just have more data" (Cleland, 2011).

Use Cases 4

Recall we introduced the four Ps: people, places, processes, and predictive analytics. The cases that follow illustrate examples where the fourth 'P' is applied to address the prior three.

Nearly all the cases below come from projects engaged with clients of the firm First Analytics. They are real cases, but due to confidentiality, many details are not shared or are disguised in some way, while trying to convey the key learning of the cases.

Some are considered to be operational or tactical in that they apply to a very specific context, such as for a specific employee during a specific shift at a specific location. Others are more strategic in the sense that they look at risk at a higher, overarching level, such as quantifying the risk factors overall or projecting trends in locations.

OPERATIONAL SAFETY ANALYTICS

Operational safety analytics refers to tools and risk scoring models that can be used in a very specific context. The predominant example is that of understanding the risk for a specifically named employee coming on shift that day.

This granular focus is the "holy grail" for some safety professionals and managers who have stewardship over teams. In reality, having the requisite data and predictive tools is uncommon. This does not mean that risk identification and quantification is not possible. In subsequent use cases that follow, examples of more overarching risk models are shown, which can give context and direction to daily operational situations.

Furthermore, some industries and companies chose to focus more on strategic factors rather than trying to get down to the location/employee shift level. This is true of the oil and gas industry, where the emphasis is on process safety.

 DOI: 10.1201/9781003364153-4

Employee Shift Risk Dashboards

Employee risk dashboards answer the question, "Who is at risk on this shift?" They provide managers with a risk score for each employee scheduled to work that day. They also provide some of the factors contributing to that score.

Dashboards can be web-based, delivered through a BI tool such as PowerBI, Tableau, or Dash, or through a mobile app. An example of a dashboard is shown in Figure 4.1.

Though the data, risk scores and factors shown in the illustration are from a real-world implementation, all but the probability scores have been changed to protect confidentiality of the company using this dashboard.

Suppose you are the manager of the Springfield Plant that produces a food product. There are three production lines, each more than a football field in length, with stations such as ingredient mixers, ovens, cooling tunnels, chillers, aligners, packagers, etc.

You have as part of your objectives the safety of the workers in your plant. You may even have a sign posted proudly showing the number of days without an incident. And somebody in the organization may be tracking safety metrics and making it available to you to measure your overall plant effectiveness and efficiency.

As part of your job, you are asked to have regular safety discussions, not only generally, but with individuals who may be considered to be at higher risk. If you only have 15 minutes at the start of the day, how do you best spend those 15 minutes? If that gives you enough time to talk to, say, three people, who should those three be?

Facility:	Springfield Plant						
Date:	12-Nov-18						
Shift:	Overnight						

Safety Risk Rank Ordering				Injury Risk	Factors with Largest Impact on Risk for Employee		
Employee Nbr	Employee Name	Line	Station	Probability	Factor 1	Factor 2	Factor 3
3145	Jasmine Johnson	3	Oven 2	0.00037	Shifts Worked	Task Experience	Training
2297	Hugo Sutton	1	Packaging	0.00024	Rules violated	Attendance	Task Experience
426	Javier Barnes	3	Aligner	0.00019	Recent Vacation	Discipline	Training
8784	Devin Richards	1	Chiller	0.00014	Shifts Worked	Field Test Failure	Task Experience
9922	Paulette Watson	1	Packaging	0.00009	Shifts Worked	Field Test Failure	Training
4090	Rodolfo Spencer	2	Oven 1	0.00008	Training	Shifts Worked	Attendance
26	Shelly Boyd	1	Cooling Tunnel	0.00007	Attendance	Shifts Worked	Discipline
1158	Kay Hubbard	1	Aligner	0.00006	Training	Field Test Failure	Discipline
4550	Wilma James	3	Aligner	0.00004	Shifts Worked	Attendance	Field Test Failure
7702	Jesus Howell	2	Packaging	0.00004	Shifts Worked	Training	Rules violated
3300	Rolando Lawson	2	Aligner	0.00003	Shifts Worked	Recent Vacation	Attendance
1200	Robin Wilkerson	1	Oven 2	0.00001	Discipline	Attendance	Field Test Failure

FIGURE 4.1   Employee risk dashboard

The dashboard tells us that we need to have those conversations with Jasmine Johnson, Hugo Sutton, and Javier Barnes. The employees coming on shift are rank-ordered by decreasing risk score (probability), and these three are at the top.

In looking at the dashboard, the numeric score is found in the *probability* column. This is the probability, or chance of that person being involved in a personal injury incident today. It is scaled 0 to 1, with 0 being the best, meaning that there is no chance of there being an incident. Note that the numbers are very small. This is good – fortunately, incidents are rare, and these small odds reflect that rarity. The actual numbers are not important; what is important is the rank ordering.

Along with the probability, the other key columns are the top three factors contributing to that score. Recall that we said we needed models that not only make good predictions, but models that explain what is driving those predictions. There are a variety of "black box" predictive models which may be better at predictions, but due to their opaqueness, they are not useful.

The scores and factors from the model are updated daily based on (1) the operational, personal and other contextual data inputted into the model and (2) the weights the model has determined to make the best predictions of those outputs.

For Jasmine Johnson, we see that the three top contributing factors to her score are:

1. Shifts Worked. This is a proxy for fatigue. It may be that she has worked many shifts in the last 7, 14, or 30 days. Or, perhaps she filled in on a couple of weekend shifts and is back on Monday, with no significant rest.
2. Task experience. The dashboard says she is assigned to Oven #2 on line three. The data and model may recognize that a substantial time has elapsed since she last operated an oven.
3. Training. She may not have undergone the most recent training for operating this piece of equipment.

Of course, many other potential risk factors, often dozens of them, may be behind the scores. The dashboard surfaces the top three for each employee. In Chapter 5, we describe an initial assessment process which generates a list of potential factors – sometimes hundreds of them. But typically, these risk-scoring models consider and give weight to less than a few dozen.

When you, the manager, undertake a conversation with Jasmine, you cannot simply say, "Hey Jasmine, Be careful today. You might get hurt." This is a disturbing statement and may be upsetting to Jasmine. In fact, she may feel as if you are targeting her, insinuating that she is inherently unsafe. A general philosophy of these tools is that these should encourage positive conversations, supported with facts. Wherever these tools are implemented, management has worked with labor, human resources, and legal to make sure that they are not used in a punitive manner. Scripts are often written to support the manager in undertaking the conversation, based on the factors that arise.

For example, the proper conversation may be, "Hey Jasmine, I recognize that it has been several months since you last operated an oven. Let's have a brief refresher to make sure you don't get burned."

Work Order Scoring Models

A variation of the employee-oriented model is one that assesses the risk of a particular task or set of tasks. A U.S. electrical utility, already known for their safety culture, built such a system, scoring a task, or work order, to be completed (Davenport T., 2020). The machine-learning model scores activities the field teams undertake, such as replacing an insulator or placing a pole.

Take the case of replacing a pole. It may usually only be a moderately risky task. But when done on the side of a hill, when rain is expected, and crane use is necessary, it becomes a much higher risk. The supervisor may want to wait until the rain subsides or assign a crew more experienced with the use of a crane. The model may also account for underlying common risk factors, such as time of day, day of week, the size, experience, and craft of the crew.

Operational Risk-Scoring Models for Incident Reviews

The intent of operational risk-scoring models is to prevent incidents in the first place. Dashboards identify potential risks to employees or with specific work tasks. They are forward-looking in that sense. But the model outputs can also be used to analyze past events. For example, when an incident review is conducted, the predictions and factors from a model at the time the incident occurred can be used to understand what risk potential and factors may have been known at the time. This is a way for management to understand what steps could have been taken, based on information that was available. It also informs ways to improve the model by (1) possibly uncovering information and facts that may be included in an enhancement of the model or (2) flaws in the data producing the predictions. Models are not static. In a continuous improvement sense, they should be updated frequently.

Map with Quartile Ranking of Risk Score
by Work Reporting Location Code

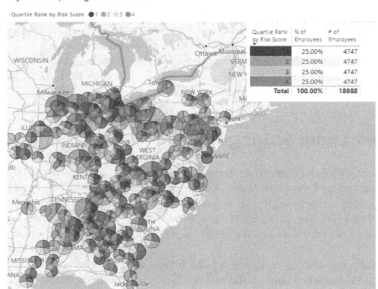

FIGURE 4.2   Map of employee risk score profiles

Geographic Mapping of Risk Profiles

Though this application of risk-scoring models is more for an overarching view of the lower-level operational data, one good way of taking a high-level view of risk is to summarize the risk scores of employees at specific locations. Since risk data is at a very low level, it can be aggregated up and summarized.

An example interactive map is shown in Figure 4.2. The risk profile of each work location is depicted in a pie chart. The size of the circle represents the number of employees in a location. The sizes and colors of the slices of the pie show the quartiles (breakouts of employees into four groups) and the magnitude for the employees in each segment. Though this figure is in black and white, in the true map, the colors range from green to red, based on the severity of the risk.

The manager's eyes are drawn to the circles with larger red segments/slices. The tool allows the user to click on a circle to reveal more detail.

TOWARD STRATEGIC SAFETY ANALYTICS: INSIGHTS ON CAUSAL FACTORS

The mapping application of risk scores is already starting to move away from tactical/operational analytics toward strategic analytics. By this, we mean gaining an understanding of the factors associated with risk overall, across employees, tasks, locations, etc. Understanding these factors helps in developing strategies such as programs, procedures, and training.

Hypothesis Development

Safety professionals and managers have long developed safety programs based on the factors that are believed to be behind safety incidents. Sometimes these are referred to as "leading indicators", although strictly speaking, at least from a data analytics standpoint, this is not the correct term. But, these so-called leading indicators are in fact considered causal factors, and we will use this broader term.

Most causal factors make intuitive sense. For example, fatigue – when a worker has been working many recent, long shifts, with little rest in between them, this worker is more likely to be involved in an incident. Thus, fatigue is a common causal factor. Because there are so many possible factors involved with safety incidents it is easy to come up with a list of many of them, derived from intuition or claimed experience in these matters. If you put operational and safety managers in a room, it does not take long to generate a list of dozens, sometimes hundreds of theorized factors. In fact, this is how a safety analytic project begins – an enumeration of any possible factor thought to be associated with increased (or decreased, as the case may be) risk.

Here is a sampling of the factors that managers typically come up with in short order:

- Being involved in a previous incident
- Experience or tenure in the position or with tasks
- Fatigue
- Lack of training
- Rules violations and discipline events
- Scores in field tests and observations
- Attendance (lateness, short-notice absences)
- Coming on or off vacation (surprisingly, managers posit both positive and negative impacts on risk)

- Operational intensity
- Variability in shift assignments (similar to fatigue)
- Short-notice call-ins
- Variety in working locations (new locations, for example)
- Drug and alcohol use
- Management attitude toward safety
- Team composition
- Weather

Human experience and judgment in developing these lists is essential and valuable. But human experience is biased and limited. First, the human brain has little ability to quantify the impact of each of these factors. Second, each individual will come up with a different, although a largely overlapping, list. The approach with safety analytics is to use a *data-driven* approach to evaluate the universe of all possible factors.

Quantifying Risk Factors

A typical safety project, at its outset, will enumerate every possible risk factor people can think of, regardless of whether it is measurable or not. The next step is to ask, "What kind of data can we get to test these factors?" This is like hypothesis testing in general scientific research; we have a theory or hypothesis that we want to either confirm or reject. The way to do that is to gather data and run formal statistical tests on those theories.

We may call this the "myth-busting" exercise. Rather than rely on human judgment and perhaps a human-based scoring of the factors, we allow the better objectivity of data and statistical models to address the hypotheses. When these exercises are complete and the results presented back to the managers, there is often surprise at the quantification and rank-ordering of the factors from the model, busting some myths.

Knowing these factors and their impacts allows us to place greater focus on things that are most likely to improve safety. Not all factors are within our control. For example, we cannot control the weather, but we can take measures if we know weather will be a factor. Or, we cannot control heavy traffic on the routes our drivers take, but we may be able to take an alternative route if a planned route appears to increase risk above a certain level. Awareness of contextual or background factors can be as useful as direct factors that we can control.

One way of visualizing factors and impacts is shown in Figure 4.3. This is based on real data from the railroad industry, although the number of factors, their names, and even the values have been changed for simplicity and confidentiality.

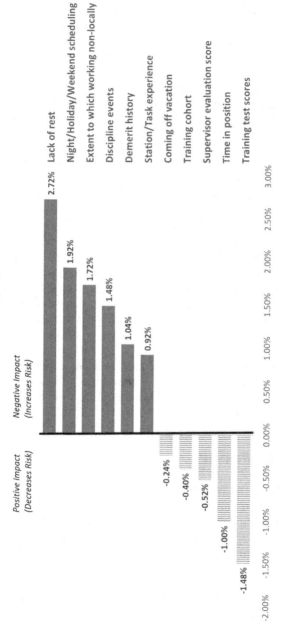

FIGURE 4.3 Causal factors and their impact on risk

This is a bar chart where the measurement is the percent increase or decrease in risk based on the presence or absence of that factor. Sometimes it is based on a one-unit change in that factor if the factor is not of the yes/no type. They are sorted, top-to-bottom, based on decreasing risk contribution. The highest-ranking risk contributors start at the top. Bars to the right of the zero-axis in the middle reflect a negative impact (increase) on risk. Bars to the left reflect a positive (decrease) impact on risk.

The top contributor to having or creating an injury incident is lack of rest, a component of fatigue. The interpretation is that, if the worker has had a small amount of rest between shifts, the risk is increased 2.72%. That may seem like a small number, but many such numbers are small when the overall risk is small. But risk factors can add up, and, in fact, in combination with others, can substantially impact risk.

These are many of the same factors that go into an operational risk application. But looking at them as high-level outputs from a model helps make decisions as to where to focus enterprise-wide efforts and investments.

The most common models used for this application are logistic regression and various versions of XGBoost. Both were discussed in the model taxonomy section. While these models are predictive, in these cases, they are also fulfilling the need to be descriptive. That is, they can explain *how* they came up with the predictions. Not all models are transparent (sometimes we call those "black box models"). In fact, XGBoost itself is not fully transparent, but the *Shapley Values* approach is one way to understand the contribution of each of the causal factors when used as inputs.

Depending on the model used, the measurement to convey the magnitude of the factor is called a *log odds ratio* in the case of logistic or other statistical probability models, or *variable importance* in the case of the more modern machine-learning methods.

Validating Models

How do we know the models work? We have discussed how intuition and judgment drive almost all safety programs today. Using the rigor of hypothesis testing and causal modeling, we can go beyond human judgment toward more objective, fact-based decision-making. We take a similar data-driven philosophy in validating this methodology. We use data to measure and assess the predictive power of our analytics approach.

The most common way to do this is to withhold data from the model (a "holdout sample" – data the model does not see) and make predictions on it from the model. We then compare the predictions to the actuals to assess the accuracy gap between the (hopefully) more accurate predictive model versus other benchmarks. The length of time needed for the holdout is usually determined by the data scientist, based on the principle of needing a large enough data sample to do a reliable evaluation.

Let's take the case of an operational risk-scoring model – one that provides a risk score for each worker shift. Though this model is operational, not strategic, we can consider evaluating the model's predictive performance, which is based on the overarching causal factors we are trying to understand, as a strategic exercise.

We take the worker risk scores and sort them from highest to lowest. At the top of the list, then, are the riskiest employees. We then work down the list by percentile, say, including up to the top 20% of risky employees. Then, we bring in the historical incident data and flag when those employees were involved in an incident on a shift. As we work down the list, we tally the incidents and can see what percentage of the actual incidents were "captured" by the model.

Figure 4.4 shows how to compare the predictive model against a benchmark or baseline. This real data is from an actual safety analytics project that focused on traumatic injuries. In this case, there are three lines, one for the model, one for the (random) benchmark, and a third for an incumbent, human subjectively weighted risk-scoring system. This allows us not only to see the improvement that a predictive model brings generally, but also to see the comparison to a legacy, subjective system.

Moving toward the right on the X-axis is equivalent to working down the list of riskiest employees, starting with the highest. The Y-axis measures the percent of correctly identified ("captured") injuries within that percentile of employees.

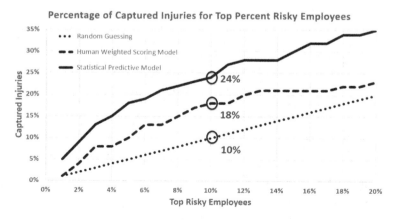

FIGURE 4.4   "Lift" curve to evaluate model performance

The dotted line represents random guessing. In other words, if we were to predict incidents simply using a random process, we would expect that process to get it right some of the time. If it's truly a random process, the percentage of injuries captured will be equal to the depth of the list of risky employees we go through. This is why it shows as a straight line, with the percentage of risky employees (X-axis) equal to the captured injuries percentage (Y-axis). A callout on the chart is at the 10% point in the X-axis. The circle on the dotted line shows that, for the top 10% of risky employees, we expect 10% of their injuries to be captured, using the random method.

We then can see how both the human-weighted scoring model and our statistical predictive model compare in terms of accuracy. At the same 10% depth, the incumbent scoring model captures 18%, versus the 10% random model. This is actually very good for a subjective weighting model, as a lot of care and consideration went into constructing it. Often, such models barely perform better than random guessing.

At the 10% riskiness cutoff, the statistical model captures 24% of the actual incidents, surpassing the legacy model. This comparison is often characterized as a "lift" over the baseline. Our lift here is 2.4, or 24% divided by 10%.

Some observations about these kinds of charts and lift:

- The area between the curve and the random line is a visual representation of the improvement that a model provides. This chart is useful in comparing alternative models. We want the rise to be quick and the gap large versus random guessing.
- The best lift usually occurs at the top risk percentages. In the figure shown, the lift starts out at 5.0, and is in the 3.0 to 4.0 range, up to the 7% cutoff. Knowing the lift factors at each potential risk cutoff point informs implementation decisions, such as how far down the risk score application should a manager go to have one-on-one risk-mitigation conversations.
- Though this does not happen often, if a model is slow to rise and increases the gap later, this is sometimes indicative of some unaccounted-for factor in the model.
- General ranges of lift in safety analytics are about 2x to 3x, with up to 4x being exceptional. While these numbers may not seem large, recall that safety incidents are very rare to begin with, and the ability to predict them is very challenging versus other predictive modeling applications. Even doing slightly better than random is an improvement and could mean that at least one employee returns home safely after work instead of getting hurt.

- Lifts can be translated into estimates of "incidents avoided". Those numeric estimates can be used in calculations involving lost work hours, costs related to the incident, etc.

Validating models, especially using visualizations like these, is a way to obtain buy-in from stakeholders as the system is deployed. Change management is an issue for safety analytics and will be discussed in a later chapter. The most successful adoptions occur when these concepts and visualizations are introduced to all stakeholders at the outset of a project.

MEASURING PROGRAMS, PROCESSES AND TRAINING

To improve safety performance, organizations are constantly evaluating and adjusting processes and programs, and creating new training. Analytics tools can inform decisions by bringing data-based recommendations as shown in the following three examples.

Impact of a New Process

Safety managers are curious as to the impact or effectiveness of new safety programs or changes in processes. But quantifying these, especially with raw data, is very difficult and potentially misleading (both positively and negatively). Models can help as they are well-suited for parsing out the effects of different potential factors in a holistic way.

One approach – though not the only approach, but a simple one – is shown in Figure 4.5. Here, the company, a tire manufacturer, implemented a process change targeting safety beginning in the first quarter of 2017. The model estimates a relationship over time to the reportable injury rate. The model is made aware of the timing of the process change event and quantifies its impact.

Then, through simulation, we project what the reportable rate would have been had the process change not been implemented. In a "what if" sense, we can compare the difference between actual (the solid line) and what the model would have expected had there been no change (the dotted line). The gap between the lines is the impact.

The bar chart on the bottom half shows the monthly percentage of improvement in the metric, as reflected by the gap in the upper chart.

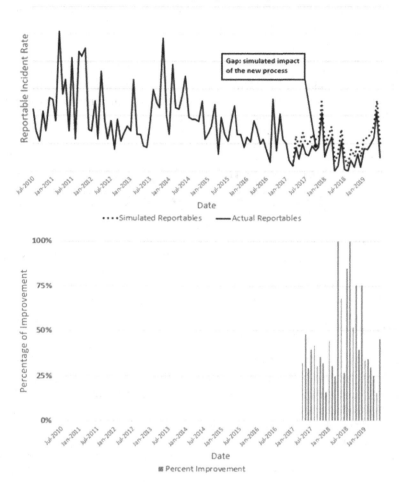

FIGURE 4.5   Impact of a new process

Measuring the Results of a Training Program

A company with nearly 20,000 field workers in 12 operating locations gradually rolled out a new safety training program. They had invested heavily in the program and wanted to measure its impact. They measured the results based on a comparison of raw incident data. Since it was a gradual rollout, the comparison was made between units that adopted the program versus those that had not yet done so.

The initial analysis appeared to show an impressive 61% decline in incidents. While they were thrilled about the prospects of the impact being so large, they had suspicions that the true number was not so large. They discovered the perils in relying on raw data alone, without accounting for other factors that could be driving the differences.

In a case like this, we can take the approach used by pharmaceutical companies when testing the efficacy of a newly developed drug. They undertake "clinical trials" where the drug is given to one group (the test group) and a placebo given to another (the control group). The principles of clinical trials are:

- Patients are randomly assigned to a test (investigational) group or to a control group.
- Care is taken to assure that the patient profiles and attributes are as similar as possible in terms of their attributes. This helps reduce any up-front biases that may cause potential differences in the results.
- The test group gets the "treatment" or new drug or therapy.
- The control group gets "standard therapy".
- Statistical models are used to analyze the outcomes. The models adjust for any imbalances between the groups that are not directly related to the treatment. This puts the groups on equal footing, such that the analysis of the resulting metric can be "with all other things being equal".

In this situation, the rollout had already occurred, so there was no luxury of creating a truly randomized grouping at the outset. Rather, they had what is known as a "naturally occurring experiment". But some of those imbalances can be handled through the modeling and through a pre-post, test-versus-control set up.

A commonly used model in these situations is called an ANCOVA, which stands for an ANalysis of COVAriance. It works like this:

1. A type of linear regression model estimates the weights that each input variable has in predicting the target variable (the coefficients).
2. The input variables are called *covariates* in that they can vary with the outcome variable, but we want these to be separate from the treatment variable.
3. After the experiment is complete, differences will be apparent in the covariates between the groups. This can be what drives the differences in the target variable, often more than the treatment variable itself.

4. We then adjust the means of the covariate to be equal between the two groups.
5. We again use the model to make a prediction of the target variable with the *adjusted* means of the covariate inputs. This produces the adjusted mean of the target.
6. The difference between the groups in the mean of the target variable, now with the covariates adjusted ("all things being equal"), is attributable to the treatment, which, in this case, is the new safety program.

The covariates used to assess training accounted for some of the following factors:

- The program was rolled out to different units at different times. Sometimes they started at high season for incidents, sometimes in low season.
- There was an underlying overall slightly downward trend over time. The model accounted for that underlying trend in relation to when the program started in a unit.
- There are variations in the core functions in each operating unit, with some functions being inherently riskier. This imbalance needed to be accounted for.
- Also related to the units is the fact that the variation of employees and their skills vary. The model implicitly accounts for the variation of the makeup of employees at a location.

Figure 4.6 shows that the adjustment is downward from a 61% raw difference to a 21% adjusted difference. The interpretation of the post model results is,

> *all things being equal*, the new safety training program showed a 21% improvement versus those who did not have the training. Those who did not have the training had an estimated incident rate of 3.93, while those who had the training [had] an estimated rate of 3.12.

While the adjustment in this case is very substantial, a 21% improvement remains impressive. And, the rigor that this kind of modeling brings lends more credence to that number.

Sometimes the multiple modeling approaches can help confirm or refine the results, offering a "second opinion" coming from another angle. A second type of model used in this case also derives from healthcare: a *survival analysis model*. This comes from an approach to understand mortality due to a health condition, such as cancer or smoking. A "hazard rate", or "hazard ratio",

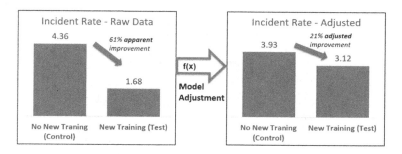

FIGURE 4.6   Raw versus model adjusted difference of a new training program

is how the model estimates the time to an event (in the case of cancer, it is death). We can project this concept, saying that the hazard rate is the increase in likelihood over time that an individual will have a safety incident.

For this model, the hazard rate becomes a function of the covariates and the treatment. Similar to a regression model, but with a specific mathematical form, the rate can accelerate or decelerate based on the influence of a covariate. And, for the treatment, the training program, we can measure how much it slows the acceleration over time of the probability for the next safety event.

The model also has attractive features, such as that some subjects will never have an event and the fact that some subjects enter or leave the study earlier or later than others (this is called censoring). Though not shown here, a comparison between the hazard rates over time between the two groups shows how the training program can reduce the likelihood of an incident as a result of receiving the training.

Evaluating Managers on Safety

Managers have safety stewardship over their employees. Their attention to safety principles and their effectiveness can vary. Companies can work with managers who need improvement and incentivize managers overall to foster a safe working environment.

But this is not easy. In this case, the safety department had hunches about who did and who did not have good safety practices. But, intervening actions based on hunches are hard to justify.

Using models like the analysis of training programs, where the model isolates the impact of the program while accounting for other factors, the analyst can take the same approach with managers. The model adjusts away any other factors it can quantify and, with a term in the model identifying the manager, can estimate his or her influence on the target metric.

Employee ID	Injury Reportable Rate Impact Index	Equipment Reportable Rate Impact Index	Weighted Average Impact
66985	1.68	1.42	1.59
98200	1.49	1.12	1.36
106995	1.00	1.46	1.16
39556	1.02	1.40	1.15
124663	1.30	1.11	1.23
187526	1.21	1.18	1.20
89663	1.25	1.10	1.20
110032	1.13	1.14	1.13
79563	1.26	0.98	1.16
100008	1.09	0.97	1.05
198665	1.01	1.02	1.01

FIGURE 4.7 Manager safety performance indices

Figure 4.7 shows a rank ordering of the top managers with the highest incident rates, as estimated by the model. But these are not incident rates themselves; rather, an index that relates their performance to the average of all managers. For example, the top manager, with employee ID 66985, has an injury reportable rate impact index of 1.68. This means that they would tend to have a rate that is 68% higher than their peers, adjusting away all other factors. What makes this possible to estimate is that managers rotate between functions and location, such that we can isolate the variation from other underlying factors.

The table shows three columns. The safety department wanted to assess managers on two types of reportable rates. The third column is a composite, simple weighted average of the two types. When sorted by this column, the safety team can see to whom they should direct their efforts. Conversely, when looking at the bottom of the list (not shown) they can analyze those managers and their styles and take what they've learned to assist other managers.

PUBLIC SAFETY

While the objectives of company safety programs primarily aim to improve employee safety, the safety of the public is a responsibility as well. Customers shop in stores where they may slip and fall, come close to electrical and gas lines, cross railroad tracks, drive on the same roads as semi-trucks, etc. Analytics can address some of these public safety concerns. Often, directing analytics toward employee safety, such as vehicle operators, helps with public safety at the same time.

Addressing Risky Highway-Rail Grade Crossings

Highway-rail grade crossings are where highways intersect with railroad tracks at grade level. There are more than 200,000 grade crossings in the United States' railroad system. About 52% of those are public crossings (Federal Railroad Administration, 2023a). A moving train can take more than a mile to stop. Most often, the motorist is responsible, due to poor judgment or behavior, for the collision. The Federal Railroad Administration (FRA) works with railroads, public road authorities and state agencies, often through regulation, to identify and improve risky crossings.

The railroad industry in general has seen continuous reductions in employee accidents and injuries. But, after decades of constant improvement, the grade crossing incident rate bottomed out in 2012 and started to trend upward again. See Figure 4.8. The rate of accidents per million train miles reached its low of 2.72 in 2012. It has been on an upward trend ever since, with the data as of when this chart was produced, through the most recent year available (2021), showing it at a rate of 3.71. A forecasting model[1] built on the historical data projects the trends out from 2021 to 2025. It shows that, if no intervention occurs, the rate is expected to reach 5.01, eliminating two and a half decades of gains.

While there is not as much a railroad can do to improve grade crossing safety, versus state agencies and road authorities, and the public themselves, they can provide direction in terms of (1) identifying the crossings most at risk and (2) recommending remedial actions to improve the safety at that crossing, both for the railroad itself and public authorities and organizations.

One major railroad undertook a project to assess all the public crossings in their network with the goal of fulfilling these two tasks. They collected data from a variety of sources, including core information from the FRA's crossing inventory database, internal engineering information, accident, and incident reports, etc.

Borrowing concepts from the insurance industry, the project team thought of risk in two aspects: frequency and severity. The goal, using data, was to

---

[1] An Unobservable Components Model (UCM), a time series decomposition and forecasting model.

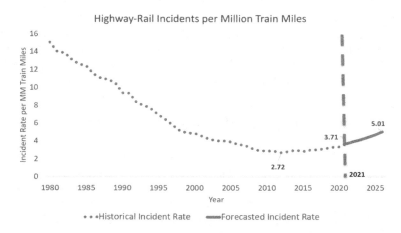

FIGURE 4.8   Highway/rail grade crossing incident rate

calculate the probabilities of having events (frequency) as well as the severity of an event (e.g., the likelihood that a bus or truck would be involved, versus a car).

One immediate challenge facing the team was that, for any individual crossing, events are quite rare. For the historical data collected, 94.9% of the crossings had no incident at all. Nearly all the remaining crossings had only one or two events, with a very small number of outliers having three, four, or five. Contributing to the low number of incidents was the fact that, when combining all the data, it provided only four years of recent history.

Prior to joining up with consultants from First Analytics, the railroad had experimented with machine-learning techniques, specifically, a variety of neural networks. Because neural networks (1) require a lot of data and (2) are mostly opaque in terms of understanding how they produce their predictions, that method was set aside after an initial assessment when the First Analytics consultants joined the team.

The team tested a variety of models that come from the statistical analysis field versus machine-learning field. The model's mathematical formulation for predicting the frequency of events, for each crossing, needed to accommodate these aspects:

- Events are rare – slightly more than 5% of crossings ever had an incident.
- The frequency counts are discrete; they fall into buckets of 1, 2, 3, etc.

- They are very heavily skewed toward a count of 1. About one-tenth of a percent fall in to the two and greater count buckets.
- There is a high preponderance of zero event counts ("zero inflated").

Using a predictive model based on the attributes of a crossing was a departure from the traditional way of assessing crossing risk. The chief safety officer said, "Where the incidents have happened, they are likely to happen again, right? That's what we were taught" (Doerr, 2016). In contrast, this method related incidents to the crossing attributes as predictors of risk, rather than the more simplistic traditional method.

The model uses crossing attributes as predictors (or causal or independent variables) and produces a probability for each crossing. The crossings can then be sorted by descending probability to focus attention on the riskiest ones.

One way of expressing the risk is to produce a risk index for each crossing, based on its probability versus the average probability for all crossings. This is illustrated in Figure 4.9 where the top 10 crossings with the highest indices are shown.

Top ten at-risk crossings with no incidents

Dept. of Transportation Crossing ID	Risk Index
#####X	487.4
#####X	462.1
#####X	447.0
#####X	444.4
#####X	442.2
#####X	436.3
#####X	428.3
#####X	424.1
#####X	422.8
#####X	412.5

$$Risk\ Index = \frac{probability\ of\ one\ incident\ for\ this\ crossing}{average\ probablity\ of\ one\ incident\ over\ all\ crossings}$$

FIGURE 4.9 Grade crossing risk index

The U.S. Department of Transportation has given every crossing in the country an ID number. Those numbers are disguised in this table. But one can see that the top crossing has an index of 487.4. With the average being 100, this crossing has a probability of having an incident that is nearly 4.9 times higher than the average crossing.

The remarkable thing about this table is that **none** of these top 10 crossings had a single incident in the four years of history covered by the data. This is the strength of **predictive** versus simple **descriptive** analytics. The traditional method, which focuses on past events at a crossing, fits into the latter type of analytics. Here, with predictive analytics, the model pools all the data and relates incidents to attributes generally, providing the ability to predict incidents at crossing where none have occurred in the history data.

Data scientists like to have ways to validate models. However, in safety that is not necessarily the case, as it usually means somebody has gotten hurt, or there has been an accident of some sort. But the analytics team on this project almost immediately had some validation in that a crossing collision occurred in the weeks following the completion of the project. This incident "went viral" with video appearing on YouTube and various media sites. Fortunately, this train colliding with a truck that had high-centered on a crossing resulting in nobody getting hurt. But as soon as the incident was reported, the team looked up the model's scores for that crossing. It ranked within the top 15% of riskiest crossing in terms of high frequency, and the top 1% in terms of severity, which was defined as involving a truck or a bus. This crossing would not have even made this list using the traditional method. The new method had it as a high-priority crossing to address with engineering improvements, although it does take time to address them once identified by a model. In other words, improvements cannot be made as quickly in the weeks following their identification.

What types of attributes may explain risk? The following is a list of several. Not all of them may be reliable predictors – that is what the model determines. And this is not a comprehensive list. But it should convey the notion that we are looking for potential factors that may influence risk (one direction or the other) that we can compile data on.

- Tracks
  - Number of tracks
  - Daily train count
  - Train count at night
  - Train types
- Crossing Road
  - Number of lanes
  - Pavement markers

- Speed limit
- Angle of approach
- Average traffic count
- Average truck traffic
- Nearby intersecting highway
  - T-intersection at crossing road
  - Distance from tracks
  - Stop type (sign, signal)
  - Average traffic count
  - Speed limit
- Engineering
  - Type of warning device (gates, lights, bells, crossbucks)
  - Quiet zone?
  - Illumination
  - Rough crossing
  - Crossing height (high-centering)
  - Crossing width

Though this model's function is mostly operational – that is, providing the focus for the crossings that need attention – the properties of this model are such that it can quantify the contribution of the various attribute inputs. We have discussed that some machine learning models, like the neural network approach that was originally tried, do not provide much, if any, insights into the inputs. In this case, the model can produce insights such as, "crossings with crossbucks have an incident risk 5.9 times higher than those with gates, holding all other factors constant".

Reducing Controllable Moving Vehicle Incidents

Companies across many industries operate fleets of trucks and cars. Employees driving these vehicles not only encounter risks to their own safety but also pose a risk to the public. This case is like the employee-shift risk dashboard case described previously. The daily dashboard provides a risk score for those employees expected to check out a vehicle for the day.

This dashboard was implemented by a utility company in a large metropolitan area, with a lot of traffic-posing hazards. Incidents can be classified by their severity, ranging from a fatality involved (fortunately this company has had none) to a simple fender-bender. Besides collisions, they also include things like bumping into a pole when backing up, driving off from a gas pump without removing the nozzle – even a horse running into the side of a truck and knocking off a mirror was in the historical data! This is why the project focuses on incidents generally instead of the narrower definition of accidents. A historical analysis to evaluate the model showed that the top 5% highest scoring employees accounted for 16% of moving vehicle incidents. This represents a lift of 3.2 versus random guessing (see the previous discussion on lift in the validating models sections – Figure 4.10 illustrates this model's lift).

Potential variables considered by the model to be possible causal factors for risk include:

- Tenure (the model found it not to be a linear relationship, but higher in earlier and in later years);
- Dozens of variations of work schedule, including overtime within various time periods, potentially fatiguing schedules, the amount of time spent in travel versus working a work order, etc.;
- The job function;
- Vehicle utilization in various periods;
- The number of sites worked;
- The number of short-notice callouts in the past 28 days;
- Percent of time spent in various districts;
- Prior involvement in moving vehicle incidents;
- Scores on training tests;
- Seasonality (late summer months increase the risk).

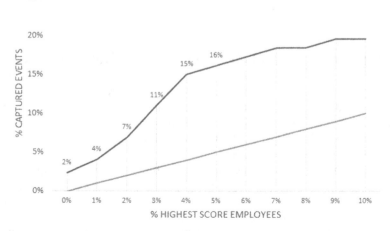

Employees with the highest propensity account for 10%~20% moving vehicle incidents

FIGURE 4.10 Model performance in identifying the employees most at risk

Targeting Human-Factor Derailments

Railroad derailments pose a potentially significant risk to the public. The Federal Railroad Administration (FRA) reports that there are nearly 1,000 derailments per year (Federal Railroad Administration, February 2023b). Fortunately, nearly all of these are minor. The events seen in the news, such as those involving the release of hazardous materials, are rare. About 27% of derailments involve human factors.

This case draws from the experiences of one of two U.S. Class I railroads that had previously implemented the employee shift risk dashboard applications described in the first use case in this chapter. Those dashboards focused on personal injury risk. It was natural to extend the application to the similar area of predicting who is most at risk of being involved in a human-factor derailment.

One of the railroads had seen a drop in derailments reported to the FRA in the previous 12 months. They wanted to continue that progress by extending the data and analytics they had built for personal injury risk scoring. In this case, the risk score was not for that day's shift, but rather for the risk that an employee would be involved in a derailment or rules violation in the next 30 days.

A machine-learning model called XGBoost (see the abbreviated taxonomy of models) considered such factors as discipline events, work schedule, rules exams, attendance and absenteeism, drug and alcohol tests, furlough, and human resource information. The project also used text mining to distill plain-text event descriptions in reports into themes and topics that could be included in the model.

A holdout analysis of the model showed that, for the top 5% riskiest employees, the model could correctly capture incidents 3.1 times better than just randomly guessing. This made better use of manager time by allowing them to focus on the most risky employees.

Though not fully attributable to the risk model, the application was a significant contributor to a 21% decline in reportables in the following two years.

DETECTING TRENDS AND SENDING ALERTS

At a higher level than looking at individual risk or causal factors, safety departments like to look at trends, keeping an eye out for alarming ones. Though this is often done through aggregate data surfaced in a report or dashboard, sometimes humans can miss picking up on a trend – especially if the reporting data is voluminous. This section describes several ways where analytical tools can augment managers by calling out trends or patterns that may escape the human eye.

Incident Rate Forecasting

This company publishes a monthly report on four key safety metrics over 26 geographies as well as for the company as a whole. Charts, like the one shown in Figure 4.11, provide a visual view of recent safety performance. But they are also forward-looking, in that they use statistical models to project the trends into the future.

There are three considerations:

- **Past Perspective**: Is there anything in the past that we missed and should better understand?
- **Current Perspective**: Did anything change last month?
- **Future Perspective**: Will these trends continue?

The future perspective, which relies on predictive analytics, is an added capability to traditional safety reports. It helps safety managers focus on what is coming, rather than on past history, which is a common practice.

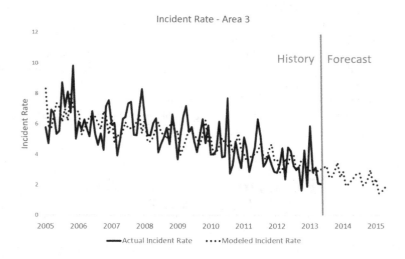

FIGURE 4.11   Incident rate forecast

Personal Injuries - Rates

Month	Incidents	Reportables	Lost Time	
Jan-2013	1.73	1.07	0.51	
Feb-2013	1.60	1.23	0.58	
Mar-2013	1.34	0.88	0.39	
Apr-2013	1.20	0.82	0.36	
May-2013	1.49	1.08	0.38	*Most recent month*
Jun-2013	1.61	1.14	0.42	
Jul-2013	1.72	1.18	0.42	
Aug-2013	1.72	1.21	0.51	*Future expected*
Sep-2013	1.62	1.09	0.42	*rates from the May*
Oct-2013	1.38	0.93	0.38	*forecast model*
Nov-2013	1.25	0.92	0.41	
Dec-2013	1.31	0.98	0.42	

FIGURE 4.12   Expected incident rates

Personal Injuries - Rates

Month	Incidents	Reportables	Lost Time
Jan-2013	1.73	1.07	0.51
Feb-2013	1.60	1.23	0.58
Mar-2013	1.34	0.88	0.39
Apr-2013	1.20	0.82	0.36
May-2013	1.49	1.08	0.38
Jun-2013	1.61	1.14	0.42
Jul-2013	1.72	1.18	0.42
Aug-2013	1.72	1.21	0.51
Sep-2013	1.62	1.09	0.42
Oct-2013	1.38	0.93	0.38
Nov-2013	1.25	0.92	0.41
Dec-2013	1.31	0.98	0.42

Actuals

1.86	1.11	0.49

June Incidents and Lost Time about 13-14%
higher than forecast, but within expected range

	Forecast	Actual	Percent Difference
Incidents	1.61	1.86	13.3%
Reportables	1.14	1.11	-2.4%
Lost Time	0.42	0.49	14.3%

FIGURE 4.13   Comparison of actual versus expected incident rates

These monthly reports showing historical data have been augmented to be predictive. Figure 4.12 shows a tabular report from May 2013. The numbers from June 2013 forward, represent the expected future rates from the model.

With the future expected rates in hand and as the actual values come in, a comparison can be made between actual and expected. The expected rates provide a benchmark as part of a variance analysis. If a rate comes in much different than expected, something has changed and needs to be investigated. See Figure 4.13 as an exhibit of this kind of analysis, conducted for the month of June 2013.

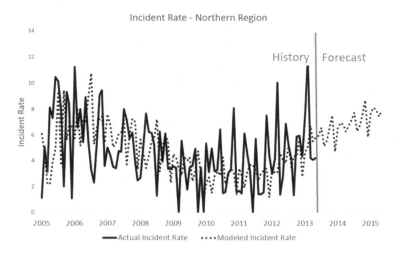

FIGURE 4.14   Forecasted incident rate

Detecting an Adverse Upward Trend

As an extension of the previous case, "Incident Rate Forecasting", these reports can be used to uncover adverse trends. See Figure 4.14 as an example, where a particular geography had a recent uptick in incidents. The model estimates this will be an ongoing increasing trend and presents estimates of the rates in future months if mitigating action is not taken.

The example cases that follow demonstrate additional extended applications of this concept, albeit with data from different companies. For example, they show how forecasts can be summarized by heat maps, or other visualizations. Additionally, a formal statistical analysis on the underlying trend is presented.

Understanding Seasonality

This company felt there was seasonality in various safety-related metrics. But this is hard to quantify, and is masked in the raw data, which has many contributing factors besides seasonality.

In this case, a sophisticated time series statistical model is used to parse out various latent factors underlying a key metric. This allows us to isolate seasonality. Here we show *Days Away and Restricted*, a subset of the *DART* measure commonly used in OSHA reporting.

Figure 4.15 shows the monthly seasonal pattern. The bars extending above the horizontal axis value of 1.00 represent a higher general underlying tendency in *Days Away and Restricted*. Note that the highest periods are the Summer and early Fall months running from July to October. Conversely, the bars below the axis represent months with lower underlying incidents. The value along the axis can be treated as a relative value, with 1.0 as the reference. Greater than 1 is higher, less than 1 is lower.

Figure 4.16 overlays the seasonal pattern bars with the actual count history, going back several years.

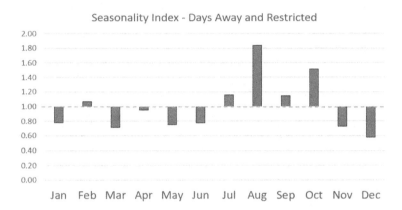

FIGURE 4.15   Seasonality index of DART

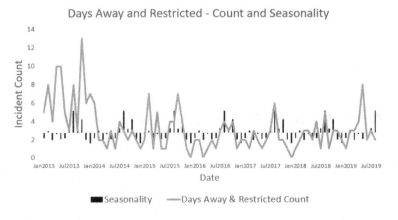

FIGURE 4.16   DART seasonality with incident count

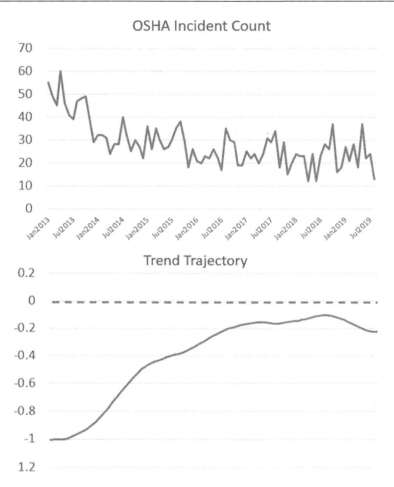

FIGURE 4.17   Incident trend trajectory evolution

Understanding the Trajectory of a Trend

This company had enjoyed a steady decrease in the *OSHA Incident Count* for several years. But the trend seemed to be slowing. In the most recent two years, visually, it appeared to be at a plateau, or even starting to increase.

Time series statistical models can isolate trends from other factors. As part of this process, they can estimate the rate of change of the trend.

In the bottom portion of Figure 4.17, which we call "trend trajectory", we see that anything below the dotted line at zero represents a still-improving trend. The more negative the value, the more the trend is continuing to decrease. When it does cross the zero-line into positive territory, that indicates that the model believes a reversal has occurred in the underlying trend.

The chart shows that the company got very close to reversing the underlying decreasing trend in 2018 and early 2019 (approaching but never crossing the zero-line) but has slightly recovered.

Drawing Attention to Business Units with Troubling Forecasts

This is an additional application of the concept shown in a previous case, *Detecting an Adverse Upward Trend*. In this situation, however, we are trying to uncover areas of concern at scale.

This company has many operating units and divisions. And their multi-dimensional business hierarchies allow for further drill downs, such as job function, location, etc., which results in many hundreds of contexts to track, at the lowest levels.

The same kind of forecasting models are applied here but operate on all the business entities simultaneously. As discussed previously, this turns insights into forward-looking ones, rather than relying on past history, which could be less reliable or outright fail to reveal anything of concern.

Figure 4.18 shows a heat map of various operating units within divisions. The eye is drawn to darker cells, which are the units within a division that are of higher concern. Though this figure is in grayscale for the purposes in this book, as implemented, it follows a stoplight formatting style. The colors range from green (no concern) to red (concern), based on the *forecasted* future expected incident count – in this case, *First Aid Incidents*.

Figure 4.19 is a secondary view, albeit with the addition of a longer (36-month) horizon, with visualizations of the expected patterns in *First Aid Incidents*.

Triggering Alerts

A popular desire for companies is to look at safety data to trigger an alert when some metric has crossed a threshold of concern. A common suggested approach is a statistical process control methodology referred to as "control charts".[2] The idea is to track a process and use information about means (averages) and variances to indicate when that process has become "out of control".

---

[2] Wikipedia: "Control Chart". https://en.wikipedia.org/wiki/Control_chart

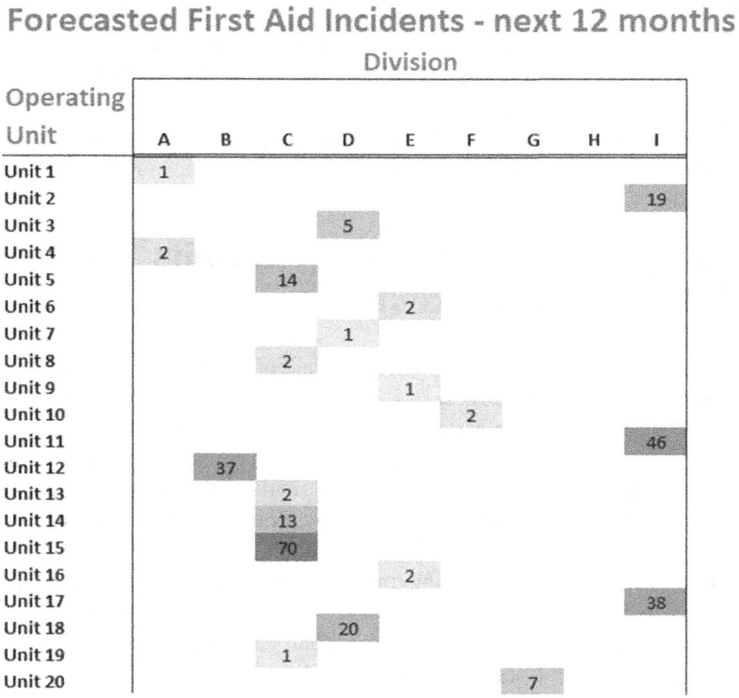

Forecasted First Aid Incidents - next 12 months

Operating Unit	A	B	C	D	E	F	G	H	I
Unit 1	1								
Unit 2									19
Unit 3				5					
Unit 4	2								
Unit 5			14						
Unit 6					2				
Unit 7				1					
Unit 8			2						
Unit 9					1				
Unit 10						2			
Unit 11									46
Unit 12		37							
Unit 13			2						
Unit 14			13						
Unit 15			70						
Unit 16					2				
Unit 17									38
Unit 18				20					
Unit 19			1						
Unit 20							7		

Division

FIGURE 4.18   Heat map of forecast first aid incidents

Conceptually, this is an appealing approach and has wide applicability in other areas of the company (e.g., six sigma quality initiatives). Applied to safety in practice, however, it suffers from several flaws.

1. Variance in safety metrics, or the spread between highs and lows, is typically large. Since these variances are used to formulate trigger points – the bound across which a metric may pass to be considered "out of control" – the resulting range can be too large and unrealistic.
2. A constant mean is assumed. Most companies across all industries in recent years have worked diligently to make safety a priority. This suggests the average has been declining over time and has not been constant.
3. In similar fashion, the variance, or spread in a particular metric, has shrunk (and thus not constant), owing to good safety mitigation measures.

Forecasted First Aid Trends by Operating Unit

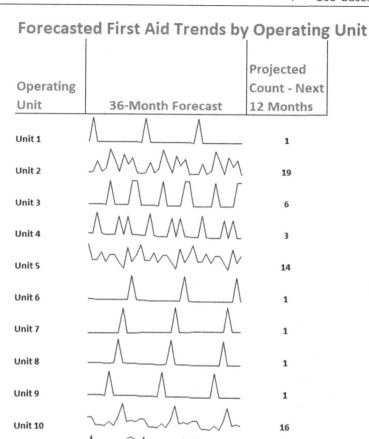

Operating Unit	36-Month Forecast	Projected Count - Next 12 Months
Unit 1		1
Unit 2		19
Unit 3		6
Unit 4		3
Unit 5		14
Unit 6		1
Unit 7		1
Unit 8		1
Unit 9		1
Unit 10		16
Unit 11		46

FIGURE 4.19 Forecasted first aid trends

Figure 4.20 illustrates these flaws. First, the "confidence levels" – the dotted lines – are too far apart, and the lower level is negative. While there was one triggering event in mid-2013, the method did not alert to any events after that. Second, one can see the decline in both the mean and the variance of the solid line (incident rate) starting in the first quarter of 2018. The pattern after that period deviates substantially from the overall mean (the blue line) and has a smaller variance.

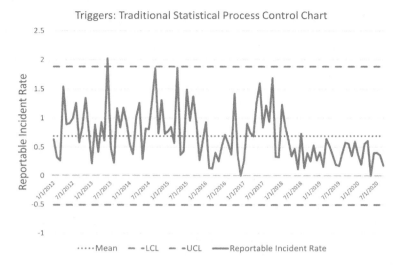

FIGURE 4.20   Traditional statistical process control chart

A better approach is to use statistical modeling and other analytical methods, which model the mean and variance as functions of time (trend, seasonality) and other factors, such as process changes. In this way, modeled means and variances – used in the same interpretive framework as control charts – can provide more realistic alerts.

Figure 4.21 shows these modeled means and variances, which adapt over time.

MACHINE VISION

Machine vision, or computer vision, is the process of training a computer algorithm to find a particular object or condition in an image. The image can be static such as a photograph but can also be video. While image processing using engineering or computer science solutions have been around for decades, since the late 2010s, recent advances in machine learning and artificial intelligence have made very significant advances in the field.

In essence, the machine vision algorithm is trained to look for something the human eye would look for. The advantage that these approaches have is that, once trained to a desirable level of accuracy, they can process images at a scale and speed that far exceeds a human's ability to do so.

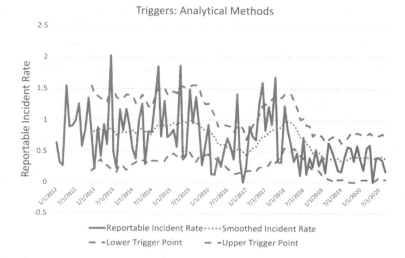

FIGURE 4.21   Analytical method process control chart

Detecting Defects and Problems in Passing Rail Cars

Several of the major United States railroads have begun to erect sophisticated video and image sensor arrays along their network.[3] These serve the purpose of train inspections. Normally, inspections are done in a rail yard, and visually, by employees who walk the length of a train looking for specific conditions that could cause a derailment or damage risk. Examples might be flaws or extreme wear in the wheels, broken, missing, or offset springs, dragging cables, etc. The cars may need a repair immediately and need to be taken out of service. Or, the defect may be noted that it needs to be resolved in an upcoming maintenance opportunity, depending on the severity of the issue. These manual inspections can take 90 to 120 minutes and involve two inspectors walking the length of the train (one on each side).

The sensor arrays offer real-time (as the train passes) inspection. They are equipped with photography, video, LIDAR, and acoustic listening sensors. The images are of extremely high resolution and high frequency. These truly

---

[3] See, for example, Union Pacific's "Machine Vision Portal" (https://www.youtube.com/watch?v=ZLHs1Fs4b2o) and CSX's "Train Inspection Portal" (https://www.youtube.com/watch?v=8mdkwb0x7-A).

are "big data" systems – a passing train can generate one terabyte of data per minute (CNBC, 2013).

A computer algorithm generally has two objectives when processing the image data. First, it must identify the object in question. Is it looking at wheels? Springs in the wheel truck? Couplers? This step of "object detection" has its own set of machine-learning algorithms that are adept at finding things in images. The goal is to isolate a subset of pixels from the entire image that we want to focus on in the next step.

The second step is to run a machine-learning classifier, typically a deep learning neural network, to identify various conditions it sees in the subset image. The classifier has been trained to look for one or more "bad" conditions.

Let's take the case of issues within the wheelset. We are looking for broken, missing, offset, or a few more types of "bad" conditions with them. Fortunately, these are very rare for a railroad. But their rareness also makes it challenging for an algorithm to reliably process. It requires many images (tens or hundreds of thousands of images, or even millions) to find enough examples to train the algorithm as to what constitutes "bad". Furthermore, those images usually are not "labeled" as such. See the discussion in the methodology chapter on the distinction between supervised and unsupervised learning. Ideally, we would like to train the computer on instances where the bad condition is identified beforehand. If they are not labeled, it becomes a much more difficult exercise. There are approaches that can work (not discussed here) that are a hybrid of supervised and unsupervised learning.

The first step is to find the spring in the image. This is not as easy as looking at a polygon defined by pixel addresses (the Xs and Ys) of the area we believe the springs can be found. Railcars come in a variety of types, including coal hoppers, tank cars, box cars, intermodal cars, etc. Each of these has their wheelsets in different locations. Even within car types, there are varying lengths and placements of the wheel sets. This makes simple approaches to finding the object to focus on unworkable. Object detection algorithms are trained to find specific objects based on showing them examples. Once it is trained, it essentially draws a box on the image showing where it found the object we asked it to look for.

The next step is to examine the springs and see if they are normal or whether one of the bad conditions exists. And if they exist, what type of "bad" is it. There are degrees of severity of "bad", ranging from "stop the train immediately, there is a derailment risk" to "examine at the next scheduled maintenance". The more severe and timely conditions are the ones we want to know about. The system will send an alert that it has found one of these to operations personnel who can issue an order as to how to resolve it.

An issue with all such systems, whether they use machine learning or not, is so-called "alert fatigue". That is, if a system is prone to issue too many false

True/False Alert Tradeoff

	Model 1	Model 2
True Positive Rate	95%	82%
False Positive Rate	14%	6%

FIGURE 4.22   Trade-off matrix for true/false alerts

alarms, the humans who monitor them tend to discount the system generally. With the advent of machine learning, this has become less of an issue, as their precision is better than legacy systems. But the issue still exists.

The way data scientists approach this is to tune their algorithms to find the right balance between capturing legitimate issues versus sending false alerts. Alas, there is a tradeoff here: One needs to find the sweet spot between the two.

To illustrate this, we use a real project whose objective was to see how machine learning might improve upon a legacy, engineering-based algorithm.[4] The results showed that machine learning could outperform the legacy system in terms of correctly identifying broken springs by a factor of 4.2. In other words, if the existing system would identify two broken springs in a sample of images, the machine-learning algorithm would find eight or nine, on average. Furthermore, machine learning was able to detect broken springs at more than twice the identification rate of humans. Of course, a machine can process images many thousands of times faster than a human.

Figure 4.22 shows the tradeoff decision regarding true positives versus false negatives. Candidate model 1 identifies broken springs 95% of the time. Model 2 identifies them 82% of the time. But model 1's better accuracy comes at the cost of issuing more false alerts, 14% of the time. Model 2, though less accurate, generates false positives at less than half the rate of model 1, only 6% of the time. Even though we like to think of data- and analytics-based systems as being completely objective, since they are based on data, there is still subjectivity involved. This is true not only in the way candidate models are selected for implementation and deployment, but also in the ways the algorithms are chosen, configured and tuned ("hyperparameter tuning").

---

[4] The legacy system was built on a variation of the Hough transform, which dates to the early 1960s. See https://en.wikipedia.org/wiki/Hough_transform

Other Machine Vision Examples

The rapid improvement in machine vision capabilities has produced a variety of applications, with varying degrees of success. Most applications are found in industries like manufacturing, mining, construction, and transportation. Examples include:

- using drone-based camera systems (unmanned aerial vehicles, or UAVs) to inspect bridges and other difficult-to-access structures, looking for maintenance issues or flaws;
- checking workers in a plant for the proper use of personal protective equipment (this applies broadly to any industry that uses PPE);
- inward-facing cab cameras to detect fatigue or inattentiveness in a driver of a truck or train;
- checking for safe fork-lift operation in a warehouse;
- detecting fires at an early stage, both in buildings and outdoors (wildfires);
- detecting when a worker enters a hazardous zone near a dangerous machine in operation;
- using locomotive-based cameras for checking railroad rights-of-way for overgrown vegetation, which creates reduced visibility for train operators and the public;
- monitoring entry into a construction site exclusion zone;
- monitoring elevated platforms on a construction site, assuring that scaffolding is being used correctly and is not over-crowded;

As this technology is still in its adolescent stage, some of these systems are not always accurate. But they have been proven to be effective in many cases. Furthermore, several issues, such as privacy and management-labor relations, have yet to catch up to the actual use and deployment of these technologies.

TEXT MINING INCIDENT REPORTS

Text mining, or text analytics, is a branch of machine learning and statistics that processes textual information with the goal of summarizing it. Often the summarizations are in a numeric form. And often those numerical summarizations are used as inputs into a more quantitative, predictive algorithm.

"Structured data" is the typical format we see with rows and columns, like a database table or a spreadsheet. Information in textual form is called "unstructured data" because its content does not easily fit into the row/column format. Typical text-mining tasks include text categorization, text clustering, concept/entity extraction, production of granular taxonomies, sentiment analysis, document summarization, and entity-relation modeling (i.e., learning relations between named entities) (Wikipedia, 2023).

Analyzing Incident Reports

The incident report or safety observation modules of safety management systems include one or more fields for providing a textual description of the incident. These fields are potentially rich in information if well-maintained. But they are a challenge for analysis in that there is no consistency in free-form inputs, several words and synonyms represent the same concept, and abbreviations are used, or there are misspellings. The concept extraction tools in text-mining software contain algorithms to deal synonyms, abbreviations, and misspellings to distill the text found in the documents into broader topics.

Much of the work in text mining is pre-processing the text for subsequent analysis. An example from a railroad incident reporting system related derailments followed these pre-processing steps.

1. Parsing
   a. Removing "stop" words: articles, pronouns, conjunctions, etc.
      i. Function words: a, an, as, for, in, of, the, and, you;
      ii. Low information words: those that appear rarely.
   b. Stem words: grouping of morphological variants
      i. Plurals: "streets" -> "street";
      ii. Verbs: "investigated" -> "investigate".
   c. Synonyms: e.g., "high-rail", "HiRail", "Hy-rail", "road-rail car"
2. Filtering: removing unwanted or extraneous words, such as "job", "work", "today".

With the basic terms and words classified into topics, the software then suggests that these are the most prominent topics found across all the documents (in alphabetical order).

- Bad order car
- Bowl hump tracks
- Bypass coupler

- Conductor/Engineer
- Crossover switch
- Excessive speed
- Fail to stop short
- Failure on the wheels
- Flag
- Fuel diesel
- Handbrakes released
- Improper line switch
- Industry tracks
- Main line
- Move light power
- Railcars
- Reverse movement
- Wye switch

After suggesting these topics, the software provides tools to explore them, including reporting on the number of terms/words that comprise the topic, a list of those words, and the ability to bring up specific documents containing those words. The tool also provides statistics regarding the prevalence of a topic, the weights given to terms within a topic, and other meaningful diagnostic statistics for use by the analyst who is building the text-mining analysis.

Text mining can be very insightful in finding underlying themes and looking for trends. An interesting application is to use the information derived from text mining in a subsequent predictive model. A study was conducted by this same railroad to see if text-mining topics could be used to improve a predictive model whose focus was to predict the cost of derailments. The traditional, quantitative factors included things like track type (main line, siding, industry), visibility, and time of day. That model had an R-squared fit of 22.8. When term indicators from the text-mining exercise such as "speed", "rest", and "broken" were entered into the model, the R-squared more than doubled, to 55.2. This is an example of how unstructured textual information can be converted into structured data and used by methods that work solely on structured data.

Augmenting and Enhancing Quantitative Data

Converting text to structured data for predictive modeling, like the previous example, need not be the overall objective. There is value in using the text-mining process to augment historical safety data that is already in a structured form. For example, records about incidents can be tagged with topics and themes distilled and classified through text mining based on the narratives found in the incident report.

In one case, a company's safety department had gone to great lengths to gather as much information about reports. This resulted in a large number of data fields that needed to be filled in, using, for example, pick lists to describe some aspect of context of the incident. The analytics team found that text mining of the narratives could largely replicate some of the fields that were being populated manually. This meant that some of the less important fields might be redundant and not necessary to use. And, since the system did not always enforce filling out the field in some cases, the text mining was able to provide more complete coverage of the concepts being captured by those fields.

Large Language Models to Summarize Incident Narratives

In late 2022, OpenAI released a model called ChatGPT. This is an AI model trained to interact in a conversational way with questions and requests. It is a type of large language model (LLM) that uses a combination of such methodologies as deep learning and reinforcement learning, with billions of nodes in its underlying structure. By early 2023, it gained high visibility, with millions of people trying it to address various topics. Some responses were eerily correct and comprehensive. Others clearly missed the mark but came with a tone of confidence despite being entirely false. This spurred a debate about the usefulness and dangers of AI generally. We won't treat those issues here but just show a few examples of how LLMs, like ChatGPT (OpenAI and used in Microsoft's Bing), Google Bard, Dolly from Databricks, and other tools, may contribute to safety analytics. These examples show the summarization of the free-form, plain text narratives of incident reports.[5]

Example 1: Narratives from human factor derailments from railroad #1.

Prompt: What can you tell me about these descriptions of derailments?

Response:

These descriptions are all related to train derailments or incidents that occurred on railways. Some of the causes include:

- *Bypassed couplers during humping operations*
- *Crew error in lining switches resulting in a run-through switch and subsequent derailment*

---

[5] These examples were generated with OpenAI's ChatGPT version 3.5 in March 2023. Due to character input limitations, only several hundred examples were provided to ChatGPT.

- *Failure to stop at a signal, resulting in a collision with another train*
- *Collision with animals (cows)*
- *Occupied car sitting on the track*
- *Locomotive fire*
- *Jammed knuckles causing a derailment*
- *Wind pushing a car out of the bowl causing a derailment*
- *Derailment caused by a foreman lining a derail under a moving train*
- *Collision with a vehicle that traveled onto the tracks*

Each incident has its own unique set of circumstances and causes, but they all resulted in the derailment of one or more cars or locomotives.

Example 2: Descriptions of operating rules violations from human factor derailments from railroad #2.

Prompt: Summarize these incident narratives.

Response:

These are a series of reports on incidents that occurred while railway employees were performing their duties as conductors or hostlers. Many of the incidents involve violations of Operating Rule 215, which relates to the proper protection of shove moves. Some incidents resulted in derailments, while others involved failure to protect a shove move, improperly protecting a shove move, directing a shove from a moving vehicle, engaging in another activity while directing a shove move, or missing a call for a train. The severity of the incidents ranges from minor to serious, and some employees received warnings, letters of reprimand, or suspensions as a result.

Example 3: Descriptions of moving vehicle (trucks of various sizes and functions) for a utility company.

Prompt: Summarize these moving vehicle accidents and incidents.

Response:

The incidents involve various types of moving vehicle accidents. Some involve rear-end collisions, such as being rear-ended while driving a company vehicle or while stopped in traffic in a personal vehicle. Some involve being struck from behind, such as driving a company vehicle and being struck from behind by another vehicle or waiting at a traffic light when the vehicle was struck. Other incidents involve incidents while backing up or making turns, and some involve being struck by stationary objects, such as a concrete barrier or a parked vehicle. There are also incidents where the cause is not specified, and incidents where the vehicle was struck while parked.

These prior three cases are examples where the LLM can produce a quick, initial summary, to provide information about the overall content of the incident reports. As a next step, we may want to get more specific information, such as a count of the types of incidents.

Example 4: Categorization of moving vehicle (trucks of various sizes and functions) incidents for a utility company.

Prompt: Categorize these moving vehicle incidents with a count of each type.

Response:

After reviewing the list of moving vehicle incidents, I can categorize them into several types:

- *Rear-End Collisions: 21 incidents*
- *Struck While Standing: 2 incidents*
- *Loss of Control: 2 incidents*
- *Struck by Another Vehicle: 2 incidents*
- *Struck on Side: 1 incident*
- *Hit Object While Driving: 1 incident*
- *Hit Object While Parked: 2 incidents*
- *Minor Contact/Fender Bender: 2 incidents*
- *Backing-up Incidents: 22 incidents*
- *Other Incidents/Unknown: 10 incidents*

It should be noted that some incidents could potentially fit into multiple categories. For example, an incident involving a loss of control could result in a rear-end collision or striking an object while driving. Additionally, incidents labeled as "Other/Unknown" may have additional details or circumstances that could further categorize them, but based on the information given, they do not fit neatly into any of the other categories.

Interestingly, ChatGPT provided some commentary at the end about some of the challenges of doing this classification. And in fact, it did not get it completely correct. But, as an initial pass at the data, ChatGPT can be useful to suggest themes to explore.

In their current state, LLMs are yet immature. They can provide conflicting answers, even when answering an identical query. They can be wildly incorrect, but when conveyed with a tone of confidence, they can seem deceivingly true. There are dangers in using this tool, and they should be used cautiously. But they can be directional and prompt ideas about investigations to undertake using other analytical tools.

Where to Go From Here

5

THE SAFETY ANALYTICS MATURITY MODEL

Many companies have a continuous-improvement function in their organization. This is typically engineering-based, where processes and systems are examined to see where improvements can be made and efficiencies gained. Applying the theme of continuous improvement to safety is helpful. What processes can be improved to result in lower risk? This is not to suggest that an approach like six sigma be applied. But it is useful to think of an organization's maturity level when it comes to safety and the improvements that can be made.

Figure 5.1 presents a safety analytics maturity model. It characterizes the maturity of an organization across four areas and through five stages.

The areas are explained below.

- Decisions and Resource Planning. This represents the extent to which evidence-based decisions are made and how those impact budgets, programs, and processes.
- Risk Analysis. The breadth at which risk is considered and assessed. At the highest level, risk is assessed at near real-time and in all situations.
- Data on Potential Risk Indicators. The amount, variety, and history of data a company collects, both internal and external, to perform safety analysis.
- Skills and Tools. The competencies of staff to work with data and analytics and the tools at their disposal to do so.

DOI: 10.1201/9781003364153-5

Stages	1	2	3	4	5
Decisions & Resource Planning	• Not particularly data-driven	• Program analysis: casual, only accounts for 1 or 2 factors	• People: evidence-based allocation of time	• Formal controlled experiments • Places: installations, types of equipment	• Evidence-based analysis shapes most budgets, programs, and processes • All appropriate dimensions (individual/team, instructor/program, task/entire process)
Risk Analysis	• Rare or none • Little or no data used	• Single focal point: e.g. employees • "Gut", anecdotal • Infrequent • Backward-looking	• Employees, public, business partners • Evidence-based • Multi-factor • Myth busting • Predictive	• Broad adoption • Field feedback • Continuous improvement • More subtle indicators	• More real time: risk scoring, alerts, feedback, and learning • Everything happens closer to "the edge"
Data on Potential Risk Indicators	• Limited • Only capture event history: Injuries Accidents Near misses	• Events • Inspection, compliance, testing	• Core "critical mass" of internal data • Some external data (e.g. weather)	• Expanded core data • Plan to capture new sources	• Tapping into non-traditional data sources (images, motion sensors, etc.)
Skills & Tools	• Excel • Std. databases (SQL) • Static reports	• Interactive dashboards • Query/drill down • Moving toward alerts	• Rigorous statistical analysis • Statistical estimates & confidence • Predictive alerts	• Text mining • Other unstructured data	• AI, Deep learning • Machine learning • Image/video analysis • Streaming data

FIGURE 5.1 The Safety Maturity Model

All companies aspire to be at the highest level, stage 5, across all four areas. The reality is that most companies can be found either in stage 3 or 2, and perhaps in one of the areas, in stage 1. Even the most sophisticated, well-resourced companies find that they fall short of stage 5 in one or more areas. As an example, a maturity assessment was performed for a large US-based oil and gas company with an excellent safety record. They were found to score at stage 5 in Decisions and Resource Planning and stage 4 in Risk Analysis. And as a data-driven organization, they were borderline between 3 and 4 on Data on Potential Risk Factors. But for Skills and Tools, they were in stage 2.

It is typical for most organizations, which strive to excel in safety, for some of their processes are very mature, but they lag in using data and analytics to support those processes. One can see that the concepts covered in this book can significantly aid a company in progressing particularly in the Skills and Tools area.

The intent of this book is to inspire safety professionals to move decisively to stage 5. The following dialogue will relate mostly to the Skills and Tools area.

Moving to stage 5, where tools like AI and machine learning are in place, is not a quick transition. And much of the ability to progress depends on the data available. But a key to doing this is developing the skills and competencies to perform analytics tasks.

For some organizations, this means building a data science team and hiring data scientists. But this is easier said than done. Here are some of the issues.

TOPIC	ISSUES
"I need to hire a data scientist, and I'm good to go."	What kind of data scientist? Can you find one with experience in safety analytics? This is like saying, "I need a doctor." Well, what kind of doctor? A cardiologist? Oncologist? Endocrinologist? While there are some good generalist data scientists available, you will make progress quicker with one that has specific experience in safety. And these are hard to come by.
Data scientists are hard to find.	This has been true for a long time and will likely continue to be for some time. A search for a data scientist can easily take four to six months or longer. Data scientists on the market have many options to choose from and can command high salaries.

TOPIC	ISSUES
"Citizen" data scientists	To address the recruiting issue, sometimes existing employees in adjacent roles, and who have a knack for data and analytics, can be pressed into service. This can work out well, but sometimes it does not. And it takes time to get up to speed through on-the-job training and coursework.
Data scientists need peers to collaborate with.	Because of the cost of data scientists, many organizations have only one, or a small number of them. But data scientists need peers to exchange ideas with. Feeling isolated leads to less productivity and turnover.
Data scientists need new challenges regularly.	They tend to have short attention spans. If they are not able to complete an assignment and move on to another, you will lose them to another organization that can provide them with that new opportunity.
Data scientists need tools.	As Winston Churchill said, "Give us the tools, and we will finish the job". Data scientists need the tools (software packages and frameworks) necessary to perform their craft and will feel unsupported if those are not provided.

The intent of listing these issues is not to scare managers away from building a data science team. It suggests that a hiring strategy needs to be informed by these issues. Additionally, the data science concepts introduced in this book should help with vetting of candidates.

Often, most of these issues can be addressed by engaging outside consulting firms. Although this comes at a cost, the benefits of doing so include:

- Speed to solution. Leveraging the experience and added capacity results in a quicker deployed solution.
- Augmentation of the data science team. Many of the engagements with outside firms are viewed as an expansion of the internal resources, thus providing additional capacity.
- Development of skills and competencies. The organization learns from the experience brought by the outside firm through knowledge transfer.

- Quality and reliability. Outside firms have experience in deploying these types of solutions in the past and know how to deploy robust, dependable solutions.
- Avoid rabbit holes. Outside firms have the cuts and bruises from going down unproductive paths and can help prevent the internal team from doing the same.

DATA READINESS ASSESSMENT

Most companies are not prepared to begin right away with predictive safety analytics. While this is largely due to process, organizational, and cultural issues, perhaps the largest impediment is not having sufficient data to begin. This is something that needs to be addressed at the very outset of an initiative when one wants progress in the safety maturity model. This is because many foreseen applications will require a number of months, or even years, of data collection to support the aspired use cases.

At the outset, it is wise to conduct a data readiness assessment. This is where a facilitator, with experience in safety analytics, leads a series of discovery discussions. It is often conducted as a partial or full-day workshop, with follow-up dialogue or additional workshops. Discussions are held with relevant subject matter experts, including professionals from EHS, IT, and operational managers. The outcome is an appraisal of where the company stands with respect to its data and its sufficiency to support advanced analytics.

Note that this is not a formal gap analysis with a roadmap as a deliverable. That could come in a subsequent phase. But this is a quick way to calibrate the starting point, to understand what plans need to be put in place. Conceptually, it is like pre-qualifying for a residential mortgage before going through the more involved steps of credit checks, income, and employment history, existing savings, etc.

Often the outcome of an assessment is that some of the most aspirational applications, such as worker/shift prediction scores, are deemed to not be immediately feasible. Nevertheless, there are usually some applications that can be identified as being feasible soon. One benefit of doing this assessment is that an organization learns more about its data than it understood before.

What follows is a sample agenda for a data readiness assessment exercise. Through "discovery", the data scientists and data engineers who know how to build predictive safety applications get a feel for the suitability of the data for the envisioned applications. This is a general agenda, with some sample questions. The actual agenda is adapted to the company, its industry, and its objectives with safety analytics. When this sample agenda is provided in advance, it spurs ideas about related topics and data sources to be included in the discussion.

DISCUSSION AREA	TOPICS	SAMPLE QUESTIONS
Company Overview	• Business operations • Employee profiles • Facilities, fleet	• Describe your business. • How many employees? Contractors? • How many facilities? • Do you focus on public safety as well as employees' safety? • Have there been acquisitions that result in process and systems integration issues? • Organization: Where does EHS fit? • Do you have a culture of analytics-based decisions? • Do you have an analytics or data science team?
Safety Culture	• Metrics • Incident history • Safety programs • Challenges	• What metrics do you focus on (e.g., reportable injury rate)? • What do trends in these metrics look like? • What reporting systems are in place for the EHS department as well as executives? • What programs have you implemented or are planning (e.g., "Total Safety Culture")? • Are field managers evaluated on safety? • Is there executive support for EHS? Is it part of a broader sustainability program within the company? • What are your top challenges and objectives today?

(Continued)

DISCUSSION AREA	TOPICS	SAMPLE QUESTIONS
Data	Enumerate your sources. Do NOT limit it to safety-specific sources, such as incident reports or compliance data. Consider what may be available from human resources, operations, etc.	• What kind of incident data is available? Examples: • Accidents, injuries, near misses, rules violations • Event classification, severity • Context around the incident • What kind of employee level data is available? Examples: • Demographics • Training • Prior incidents • Discipline • Performance reviews • Drug/Alcohol tests • Fatigue scores • Attendance • Vacation/Rest • What kind of operational data is available? Examples: • Employee assigned shift • Production line, facility, or vehicle (the context) • Task assigned • Team membership • What kind of corporate-level data is available? Examples: • Safety programs and initiatives – timing, locations, business units, employee types • Rules and policies • Legal • Employee surveys • What kind of facility data is available? • Site characteristics • Maintenance and changeouts • Process changes • Scheduled/unscheduled downtimes

DISCUSSION AREA	TOPICS	SAMPLE QUESTIONS
		• Machine historian data (e.g., PLCs) • Managers • Process changes • Where do these data reside and in what formats? • Can you show us database schemas? We are interested in the kinds of metrics (columns), the level of aggregation, the length of history, etc. • Can you show us sample rows of the primary data sources? • What future plans do you have for the addition or integration of new data or for the better use of existing data?

TAKING THE FIRST STEPS WITH DATA YOU ALREADY HAVE

It is easy to become overwhelmed with the data requirements for various predictive safety analytics applications. With respect to the data generated and needed to support business operations generally, the domain of safety analytics is among the most complex, more so than many other company processes, such demand planning.

A roadmap may indicate a data building/maturation phase that lasts upward of three or four years. But that does not mean a company cannot realize the benefits of some kind of predictive analytics before that. All the six use cases portrayed previously in the section "Detecting Trends and Sending Alerts" can be executed with spreadsheet-sized data. In other words, thousands of rows and a handful of columns, versus large database systems with many tables, millions of rows, and varying types of variables. And several of the other use cases presented can sometimes be done in a scaled-down sense.

The benefits of starting immediately with the data you already have include:

- uncovering informative, and sometimes actionable insights that were previously unsupported hunches, or unknown facts;
- informing early decisions, based on real experience with data, as to the longer-term build out of data to support safety analytics;
- beginning to instill a culture of data- and model-based decision-making.

WHAT AN IMPLEMENTATION ROADMAP LOOKS LIKE

A roadmap is a plan, mostly technology-centric, to deploy a solution. There is no one-size-fits-all when it comes to implementation. Roadmaps vary based on the type of application, the magnitude of the project, the preferred project management methodology used by the IT department, the specific tasks at hand, the length of the timeline, the kinds of resources available, and, ultimately, the solution's objectives.

Figure 5.2 presents an example of a high-level project plan and its associated steps. Of course, a more detailed plan is derived from this high-level view. Activities happen in phases, starting from design phases (solution architecture and project management), data phases (data integration and feature engineering), modeling and insights, and, finally, deployment. This diagram is oriented toward the activities of the data science and data engineering team. Some sample activities are listed for each phase. But other participants, such as IT or the business team, will have their own set of activities.

In the early steps of solution architecture and project management, the data scientists act in an advisory role, providing input into, for example, developing the business requirements or establishing the project charter, roadmap, and resource plans. Ultimately, the data science team takes the lead on the data integration, feature engineering, and modeling steps, and leads or collaborates on the insights step. Then finally, the data science team supports deployment.

A simple Gantt-like chart showing the steps of a typical implementation project, as depicted in Figure 5.2. The top, going left to right, shows the week number, from 1 to 24. Each activity is shown as a bar going for several of the weeks. There is some overlap of the activity bars.

Figure 5.3 shows how the timing of these steps play out. Note that activities overlap. Such projects are never sequential in its tasks and activities. This

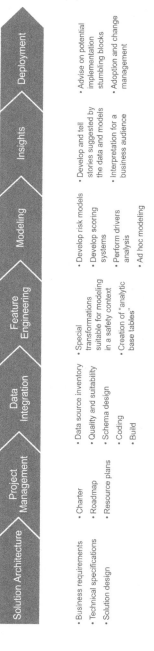

FIGURE 5.2 Typical high-level project plan

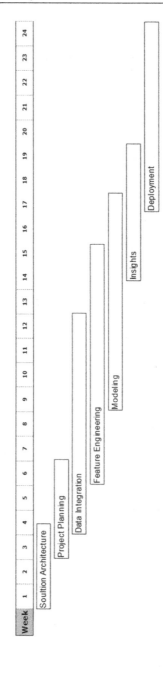

FIGURE 5.3 Timeline of project activities

makes coordination by a skilled project manager important. Furthermore, there is a feedback loop in some of the tasks. For examples, work with feature engineering may uncover errors in, or better ways of doing data integration. And modeling activities always suggest ways of setting up feature engineering.

Not shown in these process charts are proofs-of-concepts or pilots. Most projects include a phase where real data is used in a testing sense. Sometimes people use the terms interchangeably, but, generally, a proof-of-concept uses data and models to validate model performance. This is done by holding out a sample of data the model has not seen, making predictions about that data, and comparing actuals versus predicted. Pilots, on the other hand, implement the solution in the real world, albeit in a limited, trial sense.

Figure 5.4 show steps of how to undertake a proof of concept ("POC"). The concepts in the diagram overlap with the steps shown in Figure 5.2, in terms of the concepts and activities. Every project is different in how they are implemented, but POCs are generally useful to demonstrate feasibility, value, and to secure management buy-in.

CULTURAL AND CHANGE MANAGEMENT ISSUES

Most data science projects, across all domains, not just safety, end up failing. This is a disturbing fact. The reasons are many, including poor data quality (see prior discussion on data assessments), unreasonable or unclear objectives or expectations, poor project management, and poor budgeting. But a common cause relates to company preparedness to adopt the new solution. An implementation may be 100% technically sound but may never gain adoption due to cultural resistance (usually passive, but sometimes overt).

Change management is the process whereby organizational culture, practices, policies, and structures are evolved and prepared to adopt a new solution and associated processes. Particularly in older, established industries, with managers and workers with many years of tenure, these changes can be extremely difficult to put into effect. "We've been doing it this way for decades, and it works just fine" is a common rejoinder.

Larger companies today may have a change management group or a continuous improvement function. More likely, a firm in need of such services contracts with an outside party, such as one of the large management consulting firms or smaller firms focused exclusively on change management.

FIGURE 5.4   Example proof-of-concept stages and activities

Regardless of the mechanism, and despite the cost, investment in change management activities is imperative to avoid losing the effort and cost investment in developing the solution.

These considerations need to be made before the beginning of the project. At that stage, and throughout, this is a selling/marketing process. Buy-in is needed from all stakeholders, not only at the outset, but throughout the project as well. This includes:

- C-level and senior management who, by virtue of their power, can give emphasis and set objectives for an initiative;
- workers;
- managers and supervisors;
- labor unions and the company's labor relations department;
- human resources;
- the legal department;
- information technology group.

Knowledge from several successful predictive safety analytics projects demonstrated the following change management best practices.

- Buy-in is established with the COO first (and in one case, the CEO). Once it is clear that it is the COO's objective to be successful, others tend to fall in line.
- Safety-based objectives are built into individual manager performance evaluations, or, if already there, enhanced using criteria based on the new system.
- Workers are introduced to the initiative in the early stages. It is important to communicate that the system is not punitive; rather that it reflects the sincere desire of the company to watch out for their wellbeing.
- Sign-off and continual updates with the labor relations department, who, in turn, work with the labor unions.
- Solicitation from selected workers and operational managers as to hypothesized causal factors of incidents.
- Input from managers into the functionality and appearance of the end-user application.
- Reaction from managers regarding the outputs of the early preliminary models and the data sources used to build them.
- Regular monthly updates with managers and other stakeholders.
- Regular updates with senior executives.

- Approvals from human resources, especially regarding the use of private and sensitive personal information.
- Approval from the legal department on employment law issues.
- The training department was brought in early to develop training materials for the new system.
- The information technology group was given a clear understanding of the goals of the project so that appropriate resourcing could be established.
- Subject matter experts were involved in creating scripts to help managers interpret risk scores to communicate with employees, in a positive, constructive manner, who score high on a risk scale.

The companies from which these best practices derived were among some of the oldest industries, with long-established processes and with a powerful, represented workforce. Change management is not easy, but perhaps as much effort and energy may go into this overarching aspect of the project as data wrangling and data science.

We might learn from the experience of Southern California Edison (SCE), an electrical utility. They have been public with their story so we can cite what they have shared here.

Key to the success of the SCE approach has been the small, experienced, and integrated team that is addressing safety AI. Two of the key members of the team were Jeff Moore and Rosemary Perez. Moore was a data scientist who worked in an IT function (he now works in health-care IT); Perez worked in safety, security, and business resiliency as a predictive analytics advisor. In effect, Moore handled all the data and modeling activities on the project, and Perez, with many years of field experience at SCE, led change management activities.

Steps to manage organizational change started at the beginning of the project and persisted throughout it. One of the first objectives was to explain the model and variable insights to management. Outlining the range of possible outcomes allowed Perez and Moore to gain the support needed for a company-wide deployment. Since Perez had relationships and trust in the districts, she could introduce the project concept to field management and staff without the concern about "Why is corporate here?" Perez noted that it's important to be transparent when speaking with the teams. That trust has resulted in the district staff's willingness to listen and share their ideas on how best to deploy the model, to address missing variables and data, and to drive higher levels of adoption.

The team took all the time needed to get stakeholders engaged.

. . .

Perez also said the process of working with a district is critical. "You can't just walk into a district and disrupt their workflow for no reason," she elaborated. "They want to know your purpose and your objective. We try to connect, show transparency, and build trust that we are here to help, that we are here to observe how they mitigate risk, to share our findings, and to see how the findings might be integrated into their work practices. We hope they will help us understand the complexity they face every day."

(Davenport T., 2020)

The article (and eventual inclusion in a book) detailing the SCE experience also quoted from an operational field senior supervisor, whose district piloted the program.

When asked how the line crews react to the safety risk score, Todd commented that at first they would say, "Tell me something new. Every job is high risk." They were initially skeptical of the algorithm's value, but after Todd and others described to them how the model works (at a nontechnical level), most now do pay attention. Todd feels that they know the company wants to look out for their health and well-being. They view the safety risk score as another data point to take into consideration in how they do their work.

(Davenport T., 2020)

The book you are reading is not a treatise on organizational change management. But it warns that any organization with aspirations to apply its concepts is doomed to failure without considering this unambiguously associated issue.

CONCLUSION

Predictive safety analytics has the potential to revolutionize the way organizations approach safety management. By using data to identify potential hazards before they occur, organizations can proactively address safety issues and prevent accidents from happening. However, implementing a predictive safety analytics system requires a significant investment of time and resources, and organizations must be committed to building a culture of safety to make it successful.

We have discussed the key components of a predictive safety analytics system, including data collection, data analysis, and predictive modeling. We have also covered the different types of predictive models that can be used, as well as the challenges involved in implementing such a system.

One of the most important takeaways from this book is the importance of data quality and data governance. To build an accurate and comprehensive predictive safety analytics system, organizations must ensure that their data is reliable and consistent. They must also establish clear guidelines for how data is collected, stored, and analyzed.

Another important takeaway is the need for strong leadership and a culture of safety within the organization. Without a commitment to safety from the top down, encouraging change management and cultural shifts, a predictive safety analytics system is unlikely to be successful. Leaders must set the tone for the organization and make safety a top priority.

Finally, this book has emphasized the potential benefits of using predictive safety analytics in real-world situations. The case studies presented throughout the book demonstrate how organizations can use data to identify potential safety hazards and prevent accidents from happening. From reducing injury rates in manufacturing facilities to improving driver safety in transportation companies, predictive safety analytics has the potential to save lives and prevent injuries.

Predictive safety analytics is a powerful tool for organizations looking to improve safety in the workplace. By investing in a culture of safety and building an accurate and comprehensive predictive safety analytics system, organizations can proactively address safety issues and prevent accidents from happening. While the challenges involved in implementing such a system should not be underestimated, the potential benefits make it a worthwhile investment for any organization committed to improving safety in the workplace.

References

Bureau of Labor Statistics. (2018). *Standard Occupational Classification System.* Bureau of Labor Statistics.

Cleland, S. (October 3, 2011). Google's "Infringenovation" Secrets. *Forbes.*

CNBC. (September 12, 2013). *Train Kept a Rollin'.* Retrieved from https://www.cnbc.com/video/2013/09/12/train-kept-a-rollin.html

Davenport, T. H. (2006). Competing on Analytics. *Harvard Business Review, 84*(1), 98–107, 134.

Davenport, T. H. (July 30, 2020). Machine Learning and Organizational Change at Southern California Edison. *Forbes CIO Network.* Retrieved from https://www.forbes.com/sites/tomdavenport/2020/07/30/machine-learning-and-organiza tional-change-at-southern-california-edison/?sh=552e34a33336

Davenport, T. H., & Harris, J. G. (2017). *Competing on Analytics: Updated, with a New Introduction: The New Science of Winning.* Harvard Business Review Press.

Doerr, R. (2016). *Traffic Struck-by.* Retrieved 2023, from https://vimeo.com/528234983 2:17 mark

Federal Railroad Administration. (2023a). *Highway-Rail Grade Crossing and Trespassing Research.* U.S. Department of Transportation. Retrieved from https://railroads.dot.gov/research-development/program-areas/highway-rail-grade-cross ing/highway-rail-grade-crossing-and

Federal Railroad Administration. (February 2023b). *1.12 Ten Year Accident/Incident Overview.* FRA Office of Safety Analysis. Retrieved from https://safetydata.fra.dot.gov/OfficeofSafety/publicsite/Query/TenYearAccidentIncidentOverview.aspx

Wikipedia. (March 18, 2023). *Text Mining.* Wikipedia. Retrieved from https://en.wikipedia.org/wiki/Text_mining

Index

Printed in the United States
by Baker & Taylor Publisher Services